THE ROYAL NAVY AT MALTA

VOLUME ONE:

THE VICTORIAN ERA

1865-1906

A collection of old photographs taken by the Ellis family at Malta

Compiled by
LT CDR BEN WARLOW RN

MARITIME BOOKS

First published in 1989 by
Maritime Books, Lodge Hill,
Liskeard, Cornwall, U.K.

ISBN 0 907771 43 2

Typeset and printed in Great Britain by
Penwell Limited, Callington, Cornwall.

PREFACE

THESE photographs are from the collection of Mr R. Ellis of Malta, whose family have recorded the life and development of the Royal Navy over the past one hundred years. A selection has had to be made from the wide range of photographs available, and the aim has been to give as broad a coverage as possible within the constraints of a reasonably sized book. The captions have been prepared to place each view in context, and to add a little information, though the old adage that a picture is better than a thousand words still holds true here. Malta has a special place in the history of the Royal Navy, having been one of its main bases for over a century, and being uniquely positioned by geography athwart the sea lanes of the Mediterranean, and also on the routes to and from the East once the Suez Canal had been opened. Because of the calmer seas experienced in that area—though the seas can be violent there in the extreme at times—the Channel Fleet was often taken there to exercise, and it always proved a popular visit for the sailors, who found the island's climate and hospitality equally enjoyable. Thus a wide variety of ships were to be seen in Grand Harbour and Sliema Creek. This first book covers the period of the Victorian Navy, with an overlap into the Edwardian era, which will then form the starting point for the second volume to be published later.

HMS REVENGE

1865

THE *Revenge,* launched in 1859 at Pembroke Dock, was a 91 gun screw Second Rate Ship of the Line. Here she is seen when Flagship of Admiral Hastings R Yelverton CB, lying in Grand Harbour with Valletta in the background. She was ordered home in that year and was then employed on Coast Guard Service at Pembroke Dock. In 1872 she became the Base Ship at Queenstown, and in 1890, when lent to the Clyde Industrial Training Association, she was renamed the *Empress.* The White Ensign she is wearing had only just been authorised for wearing by all ships of the Royal Navy (on 5th August 1864) in place of the differing ensigns distinguishing the Squadron to which a ship belonged.

HMS TEMERAIRE—Ship's Galley

1885

THE men in this galley show the preference for beards which lasted from the 1870s to the 1890s, the Admiralty Circular permitting the wearing of beards being issued on 24th July 1869. The initial 'T' on the boat's bow was typical of the early distinctive designs used prior to the introduction of formal ship's badges in 1918. *Temeraire* had recommissioned in 1884, two years after the Alexandria action, and in this commission she landed a Naval Brigade which saw action at Tel-el-Kebir and at Khartoum. The fitness aspect of this photograph is appropriate as the present *Temeraire* is the Naval School for Physical and Recreational Training.

Previous pages

HMS TEMERAIRE

November 1878

LAUNCHED at Chatham in 1876, and completed in 1878, *Temeraire* sailed at once for the Mediterranean where she served for the next 14 years. She is seen here after returning from the Dardanelles, which the Squadron forced in a snowstorm. She was the first British barbette ship, and used her guns a few years later at the bombardment of Alexandria in 1882, when she silenced Fort Mex, and helped attack Fort Pharos and Fort Moncrieff. She had the longest spars ever carried (115ft). Although paid off in 1891, she was not sold until 1921, having served as the Training Ship *Indus II*, the reformatory ship *Akbar* and as a Depot Ship at Invergordon. Beyond her, in dry dock, can be seen the five masted *Agincourt* (1865), then Admiral Commerell's Flagship (see page 27)

HMS ROYAL OAK

1867

THE Ironclad Frigate *Royal Oak* was launched at Chatham on 10 September 1862, and sailed for the Mediterranean in 1863, being the first British ironclad to pass the Straits of Gibraltar. In 1864 she achieved 13½ knots under sail, a knot better than her maximum speed under steam propulsion. She was fitted with a 19ft hoisting propeller. She had square rig added in 1866, and in 1867 she was re-armed. In that year she was part of the Mediterranean Fleet but attached to the Channel Squadron for a period. She was present at the ceremonial opening of the Suez Canal on 17 November 1869. At the end of 1871 she returned to the United Kingdom and was sold on 30 September 1885.

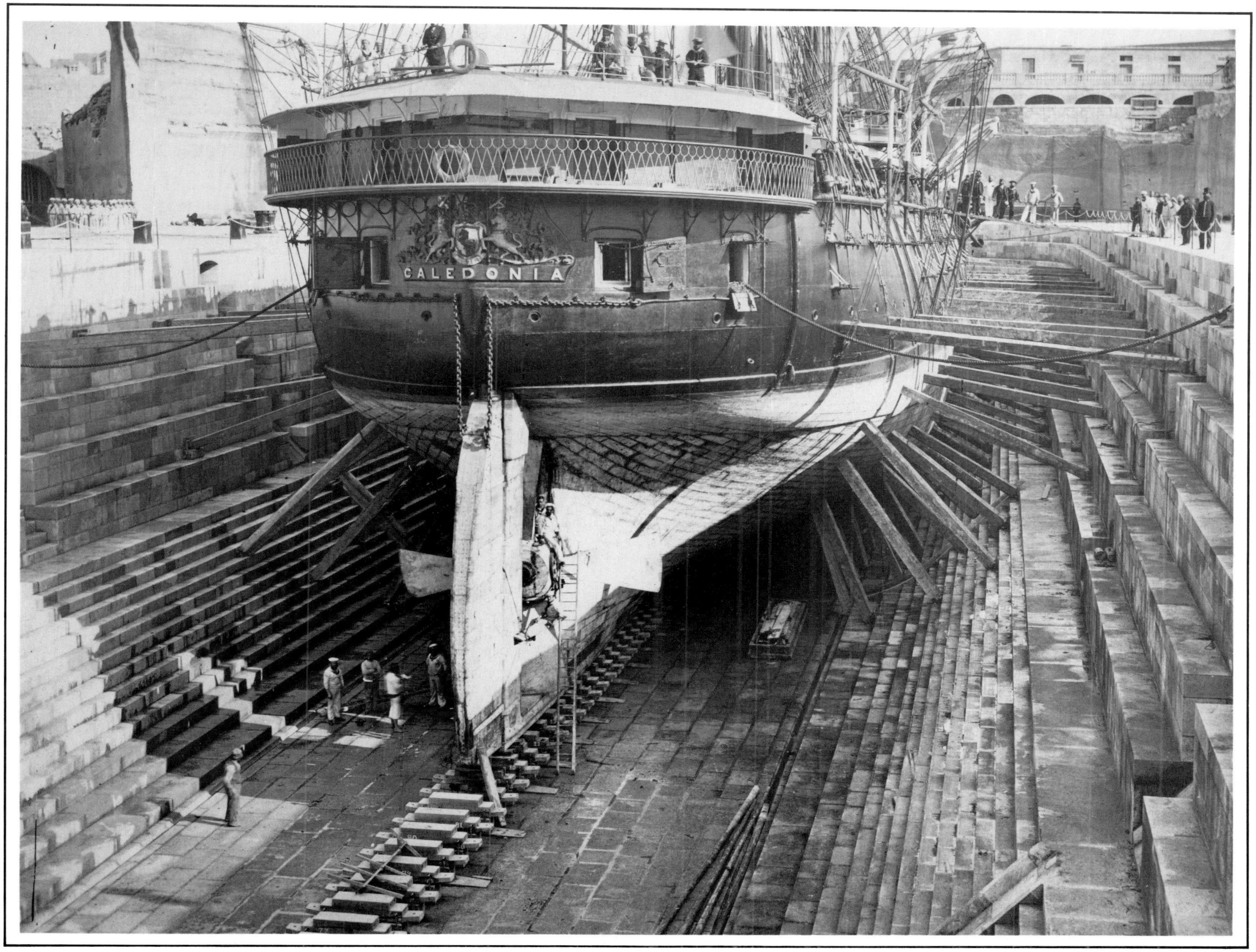

HMS CALEDONIA

LAID down as a 92 gun Second Rate at Woolwich in 1860, *Caledonia*'s conversion to an Ironclad was approved in 1861. She was launched at Woolwich on 24 October 1862, but not completed until July 1865. Her original barque rig was changed to ship rig in 1866. The weight of her machinery proved too great for her hull, which suffered from sagging as a result , entailing alterations to her shafting. Here she is seen in dry dock in French Creek, with windsails rigged to keep her messdecks cool, and only the bases of her two funnels visible. Originally she had a flush upper deck, but the poop and topgallant forecastle added later can be seen here quite clearly. When the forecastle was added the original angled iron bowsprit was replaced by this wooden, nearly horizontal one. Also visible is her large (21ft diameter) four bladed propeller which drove her at 12½ knots. On trials she reached 12.9 knots. The smartness of the armed sentries on the forecastle and jetty contrasts with the casual cheerfulness of the other members of the crew who can be seen relaxing on the upper deck. Shortly after this picture was taken (probably 1872) the *Caledonia* left the Mediterranean for the last time. She was sold to Castle of Charlton for breaking up in 1886.

HMS DECOY

1875

A COMPOSITE screw gunboat of the *Ariel* Class, *Decoy* was launched at Pembroke Dock in 1871 and was commissioned at Devonport for the North America and West Indies Station in 1872, under the command of Lieutenant and Commander John Hayes. The half stripe for Lieutenants of over eight years' seniority was introduced in October 1877 but the rank of Lieutenant Commander was not formally introduced until April 1914. *Decoy*'s crew saw service in the Naval Brigade which took part in operations on the Gold Coast against the King of Ashantee the year before this photograph was taken. She was later present at the bombardment of Alexandria in 1882, took part in the blockade of the Nile in 1883 and her crew were involved in the campaigns in Egypt and the Sudan 1884-1885. She was sold at Malta in October 1885.

HMS HERCULES

28 November 1875

THIS Central Battery ironclad was launched on 10 February 1868 at Chatham with eight 10 inch muzzle loading guns in the central battery, which was fitted with embrasures to improve the arcs of fire. She had a 9 inch gun forward and another aft. The forward gun was on a pivot, with ports in the bow and on each side, which meant her anchor cables had to come inboard on the upper deck instead of the maindeck. She also carried 7 inch guns fore and aft, being the first ship with a three calibre main armament. Just prior to this photograph being taken, she had had steam steering gear fitted. She is seen here as the flagship of the Commander-in-Chief, Mediterranean (The Hon Sir James R Drummond KCB) and she was still his flagship the next year at the Dardanelles demonstration. She was modernised in 1892, her sailing rig being removed. Later she became the depot ship at Gibraltar, being renamed *Calcutta* in 1909. In 1915 she became the Artificer training ship *Fisgard II* at Portsmouth and was sold in 1932, her hull being taken to Preston in that December for breaking up.

HMS NORTHUMBERLAND

BUILT by Mare and Co at Milwall, she had been laid down in 1861, but was delayed by design changes and finally was too heavy for launching. It took pontoons, tugs and a high tide to launch her in 1866. She first commissioned on 8 October 1868, serving in the Channel Squadron until 1875. She was then refitted, during which time she was re-armed. Her bowchasers were upgraded from 8 inch to 9 inch guns and her two 8 inch guns aft were replaced by a single 9 inch gun. The after two 7 inch guns were replaced by 8 inch guns. In the central battery, viewed here from aft looking forward, the original arrangement of two 9 inch guns and nine 8 inch guns on each side has been adjusted by the removal of an 8 inch gun from each side. The two, larger 9 inch guns can be seen here as the 6th and 7th from forward in the battery. She was a half sister to the *Minotaur* and *Agincourt* (see page 27) and could steam at 14 knots. She originally had five masts but her second and fourth masts were removed in 1875, when she became barque rigged. Her best speed under sail was 7 knots. In 1909 she became coal hulk CD 68 and she was sold in 1928, being transferred to Dakar as the hulk *Stedmound*.

HMS THETIS

THE *Thetis* was a Briton Class wooden screw corvette of 1860 tons, launched at Devonport on 26 October 1871, striking the Flagship *Royal Adelaide* in the process. She was completed in January 1873. She commissioned for the China Station the next month and transferred to the East Indies Station in that December. She is seen here in dock in Malta, probably prior to her return to Plymouth in June 1877 to pay off. Her single shaft was fitted with this Griffiths fixed propeller of 14 ft 9 inches diameter, it had detachable blades and a disconnecting tail shaft. The single striped Officer is a Mate or Sub Lieutenant, the thick stripe having been introduced for them in 1861, and Lieutenants, Commanders and Captains, then all being given an extra stripe. *Thetis* served a commission on the Pacific Station (1879-1883) and was sold in November 1887 to Pethwick at Plymouth. Her nickname was the *Tea-Chest*, earned by her predecessor in 1823 when she underwent trials in which the rum ration was halved and tea introduced into the Service.

HMS DEVASTATION

1877

THE *Devastation* was the first 'mastless turret ship', the removal of the sails giving the guns a good field of fire. She was also the first iron ship built at Portsmouth, being launched there in 1871. She first served in the Channel under Admiral Hornby, but commissioned for the Mediterranean in 1873 and remained there until the year after this photograph was taken. She then returned to Portsmouth for a refit when she was fitted with Whitehead torpedoes, Nordenfelt guns and searchlights. Her Captain in the Mediterranean spoke of her accommodation being 'wholesome and comfortable for Officers and men', but the sailors who served in her and her sister (the *Thunderer*—see page 18) called them 'rat holes with tinned air', and the 1878 refit did include improvements to ventilation and sanitary arrangements. The 'cul de sac' on the quarterdeck allowed the after 12 inch guns to be depressed, and the boats were stowed on a flying deck. Her novel low freeboard, easily visible in these photographs, was the cause of much pessimism, but she proved to be a steady gun platform and a good steamer and seaboat. Even when her forecastle was washed down, she remained dry amidships. She was sold in 1908 for £21,700, 5 per cent of her original cost.

HMS THUNDERER

16 November 1878

THE Turret Ship *Thunderer* was launched at Pembroke Dock on 25 March 1872 and commissioned at Portsmouth on 1 May 1877. She, and her sister ship *Devastation* (see page 17) were the basis of future battleship design, being the first seagoing mastless turret ships and the first seagoing ships of the 'quasi-monitor' type. They introduced the conning tower which housed the vital controls for the steering, engines and gunnery, all in one armoured space. She recommissioned on 1 October 1878 under Captain A J Chatfield, and in fine weather sailed for the Mediterranean, arriving at Malta on 4 November. Her forward 35 ton gun had been replaced by an Armstrong 12.5 inch muzzle loading gun which introduced hydraulic loading, reducing the crew from 22 to 10 men. On 2 January 1879, a few weeks after this photograph was taken, the gun was double loaded during a practice and it exploded, killing 11 men and wounding 30 others. But for this accident, muzzle loading may well have held sway in the Navy for a longer period. In 1889-1891 she was given a major refit, being fitted with new machinery, but she had to return home from the Mediterranean in 1892 with boiler trouble. In 1907 she was declared non-effective, and was sold in September 1909 to Garnham.

HM SHIPS ALEXANDRA, HECLA AND SUPERB

Early 1880s

THE *Alexandra* (foreground) is seen lying in the Grand Harbour. Her hull shows the substantial embrasures designed to permit her main guns which were concentrated in a central citadel to fire virtually ahead. She served in the Mediterranean from her completion in 1877 until 1889. In 1882 she was in action at Alexandria (see page 24) and where she fired forty eight 11 inch shells and two hundred and twenty one 10 inch shells. In 1885 she carried out the first night gunnery practice by searchlight. In 1886 as Flagship of the Duke of Edinburgh, she was painted white. Three years later she was fitted with one of the first seagoing operational wireless transmitters. The *Hecla,* seen lying beyond her, (see page 50) was a torpedo boat depot ship, purchased in 1878, and was commanded by Captain Arthur Wilson between 1881-1884, an expert in torpedoes. He developed submerged torpedo tubes in the 1890s and also a net cutting device for fitting on torpedoes. The two funnel vessel on the left is the *Superb* (1875) (purchased 1878) seen here with her funnels increased in height, carried out in 1881.

HMS INFLEXIBLE

16 October 1882

SEEN here at Malta under examination after the bombardment at Alexandria in the July where she fired eighty eight rounds of 16 inch against Forts Ras el Tin, Mex, Ada and Pharos. Her guns were the heaviest in the Service and her firing was impressive and accurate, but found not to be as destructive as first thought when the damage was inspected later. She acted as flagship during the action and was damaged when a 10 inch shell hit below the waterline outside her armour and glanced upward and penetrated her deck, killing several people. The blast from her own guns damaged her upperworks and smashed boats. She returned to Portsmouth, where she had been launched on 27 April 1876, for full rectification of defects. Her rig was removed and tops fitted to the masts. Her four 16 inch guns can be seen clearly in this view, the starboard gun barrels are pointing to the port side. She was commanded by Captain J A Fisher CB at this time. She carried the thickest armour ever used in the Navy (24 inch) and had anti-roll tanks. She was also fitted with electric light. She was sold in 1903 to Ward, Birkenhead and Preston.

HMS TAMAR

1882

THE *Tamar* was an Iron Screw Troopship of 4650 tons, launched by Samuda, Poplar, on 5 January 1863. Some of her crew were in the Naval Brigade that defeated the Ashantee at Amoaful in 1874. In the year this photograph was taken, *Tamar* travelled between Portsmouth and Malta three times and was with the force that bombarded Alexandria in the July. In the late 1890s the Government decided to replace troopships by hire for tender arrangements, and that became the future pattern until air trooping was introduced. In 1897 she became the Base Ship at Hong Kong, being scuttled there on 12 December 1941, when the Japanese invaded. Beyond her, alongside the jetty which was later to become part of *HMS St Angelo*, can be seen the *Hibernia*, a First Rate Ship of the Line, launched in 1804 and which was the Base Ship at Malta from 1855 to 1902. (see page 28).

HMS SULTAN

1889

ONE of the largest central battery ships, the *Sultan* (9290 tons) carried eight 10 inch and four 9 inch guns in a two tiered battery fitted with embrasures and an overhang to improve the arcs of fire. Completed in October 1871 at Chatham, she was detached to the Mediterranean Fleet to take part in the bombardment of Alexandria in July 1882. During later service in the Mediterranean, she grounded in the South Comino Channel, just North of the main island of Malta, on an uncharted rock on 6 March 1889. Her bilges were damaged, and efforts were made by the *Temeraire* (see page 6) to pull her off. These efforts proved unsuccessful, and she steadily settled as water continued to enter her hull. On 14 March a gale sprang up, and she slipped off the rock and sank. The Italian firm of Baghino and Co was awarded the contract to salvage her (first photograph). On 27 August she entered Malta Dockyard for temporary repairs (second photograph). In December she was able to make the passage to Portsmouth under her own steam, arriving there on 22 December. She was then placed in reserve prior to a full modernisation in 1893-6. Later she was dismantled to become a hulk, and was used as an Artificer Training Ship *(Fisgard IV)* from 1906 to 1931. Renamed *Sultan* in 1932, she was used as a Mechanical Training Ship until 1940, when she became a depot ship for minesweepers to 1945, still at Portsmouth. She was sold in 1946, being taken to Dalmuir for breaking up.

HMS ALEXANDRA

1882

ORIGINALLY to have been named *Superb,* she was launched at Chatham Dockyard on 7 April 1875, by HRH The Princess of Wales, later Queen Alexandra. She was the first British Ironclad to be launched by a member of the Royal Family. On commissioning in 1877 she went to the Mediterranean and was the Flagship of Admiral Hornby when he forced the Dardanelles in 1878, running aground on her way up in a snowstorm. She took the brunt of the Egyptian gunfire at the bombardment of Alexandria in July 1882, and some of the damage can be seen in these two photographs taken on her return to Malta. She had over 60 hits, suffering considerable damage to her upper deck, boats and rigging, and had her hull penetrated 24 times above the armoured plate. She lost one man killed and three wounded, but there was no serious damage to the ship. Her Gunner, Israel Harding, was awarded the VC for extinguishing a burning shell fuse on the deck during the action, the award being made at Malta in the November. She returned to the United Kingdom in 1889 and was sold in 1908 having spent her last few years in reserve.

HMS MONARCH

1883

LAUNCHED at Chatham on 25 May 1868, she was the first seagoing turret ship and the fastest battleship of her day at 15 knots. She was the first oceangoing warship to fight under steam instead of sail and also had the longest life on the effective list of any British armoured ship. Her main armament of four 12 inch guns was in two turrets, placed close together, rather than inside her hull. With her high freeboard, her design was more successful than that of the ill fated *Captain.* The year before this picture was taken she was at the bombardment of Alexandria, where she fired one hundred and twenty five 12 inch shells, fifty four 9 inch and twenty one 7 inch shells. Seen here as the Flagship of the Commander-in-Chief, Vice Admiral Rt Hon Sir John Hay KCB, she is cleared for action with her topmasts sent down and her bulwarks lowered to give her large guns a clear line of fire. She has some lighter guns on the flying deck above the forward turret. She was stationed at Alexandria for several years. In 1885 she was ordered to Malta during the Russian war scare, but she broke down at sea and was towed to Malta, repaired and returned to England under escort. In 1890 she was modernised and then served as Simon's Bay guardship, later becoming the Depot Ship *Simoom,* returning to the UK in 1905 and being sold the next year.

HMS AGINCOURT

April 1884

THIS five masted Iron Screw ship was built at Lairds at Birkenhead, completing in 1867. She served in the Channel before sailing for the Mediterranean in 1877 after a refit when searchlights were fitted. She was Rear Admiral Commerell's Flagship in the Dardanelles expedition of 1878. Here she lies in Grand Harbour with her flags at half mast indicating a death onboard, or in the Fleet. She is wearing the Flag of Rear Admiral J C Wilson, Second in Command of the Channel Squadron. She, with the *Achilles* and *Northumberland*, were visiting Malta and sailed for Syracuse on 14 April prior to returning to Plymouth on 12 May. A year later she joined Admiral Hornby's Particular Service Squadron. This Squadron was gathered for a Baltic Expedition, but, instead, exercised torpedo boats and tested the Fleet's defences against them. The value of these craft was well demonstrated during these exercises. She became *Boscawen II* in 1904, *Ganges II* in 1906 and coal hulk C 109 in 1908, not being broken up until 1960, 95 years after she had been laid down.

HMS HIBERNIA

A FIRST rate, launched at Plymouth Dockyard on 17 November 1804, she had spent 12 years on the stocks. She spent most of her active life as flagship of the Mediterranean Fleet, and in 1855 became the Base Flagship at Malta. The picture above, taken in July 1894, shows her afloat, but with her rigging depleted and a permanent canopy fitted over her forecastle. Her bowsprit has been removed, but her figurehead remains watching the harbour. She was still the flagship of the Admiral Superintendent (then Rear Admiral R Duckworth-King). The photograph of some of her crew shows a mixture of artisans, blacksmiths, carpenters, shipwrights, artificers and their mates. Those with badges sporting a horizontal gun barrel are armourers and their mates. Note the rating top left with a single anchor and crown. At that time there were first and second class Petty Officers, the rate of leading hand being introduced in 1907. The Petty Officer on the right of the middle row has five bars to his General Service Medal. Note also the style of ties worn by the men in fore and aft rig. The *Hibernia* was replaced by the *Achilles* in 1902 and was then sold. She was broken up at Malta, and the last photograph shows her in April 1903, during this sad process. Her name lived on, however, for on 6 January of the next year the keel was laid of a new *Hibernia*, a battleship of the *King Edward VII* Class, at Devonport, where this old *Hibernia* had also been built.

R. ELLIS

HMS HIMALAYA

April 1884

AN IRON screw troopship of 4690 tons, the *Himalaya* was launched on 24 May 1853 by Mare of Blackwall. Originally bought by the P & O Company, she was the largest steamer in the world when built. Her saloon was 100 ft long and could seat 170 people. Originally chartered, she was later purchased outright at a time when trooping was at its height. In 1855 the *Himalaya* was one of 68 troopships in Table Bay, carrying troops to and from India and China. She could steam at 14 knots, and with her sails assisting could reach 16½ knots. She gained the Battle Honours New Zealand 1863-6 and Ashantee 1873-4. Here she has just embarked the paid off crews from the *London* (2nd rate of 1840, converted to a screw ship in 1858, harbour store ship 1874 and sold at Zanzibar 1884) and the *Dryad (Vestal* Class sloop 1866 that had been on the Far Eastern Station to 1884). She sailed with them on 9 April for England, calling at Gibraltar on the way and reaching Devonport on the 18th. She became the coal hulk C 60 in December 1895 and was sold to E W Payne in September 1920. She was sunk during an air raid on Portland Harbour in 1940.

Previous pages

HMS MAGPIE

6 June 1884

THE *Magpie* arrived at Malta on 31 May, having sailed from Singapore on 22 March and travelled via Batavia, Sunda Strait, the Seychelles, Aden and the Suez Canal. A *Plover* Class screw gun vessel and sister to the *Woodlark* (see page 33), she had been launched at Portsmouth on 12 February 1868. This was the last class of British gun vessels constructed of wood. She had been employed on survey duties in Far Eastern waters, including the waters around Japan and Hong Kong. She has a charthouse amidships and a different forecastle and poop to her sister. She was under the command of Commander Hon Foley C P Vereker, who had joined her at Labuan in April 1883. He later commanded the *Rambler* (see page 34). *Magpie* left Malta on 8 June and visited Gibraltar on 17 June prior to arriving at Plymouth on 24 June 1884 to pay off—her paying off pennant can be seen on her mainmast. She was sold in September 1885.

HMS WOODLARK

1884

A WOOD screw gun vessel of the *Plover* Class, *Woodlark* was launched at Chatham on 9 March 1871 and displaced 805 tons. She was fitted with twin propellers to give her a shallow draught, but this, and a traversing stern chase gun, precluded the fitting of hoisting screws, impairing her sailing qualities. These were the smallest vessels in the Navy to have telescopic funnels. The topgallant forecastle and poop were added later to improve the accommodation in the tropics. She is seen here at Malta, having arrived on 2 February 1884, after passing through the Suez Canal from the East Indies Station, where she had been the Senior Officer's ship in the Persian Gulf. She paid off at Malta on 3 March and recommissioned on the 4th under Commander Anthony Kingscote. She then returned to the East Indies Station, reaching Suakin on the Red Sea on 15 April. In November 1885, her boats were used with the Irrawaddy Flotilla when the Burmese were defeated at Myingyan. She was sold at Bombay on 9 March 1887.

HMS CRUISER
1884

CALLED the *Cruizer* until 1859, she was a 960 ton wood screw sloop, launched at Deptford on 19 June 1852. She was with Rear Admiral Hon Richard Dundas's Squadron at the bombardment of Helsinki in August 1855. Later she became a training ship for ordinary seamen in the Mediterranean. Most of her year was spent on cruises to Sicily, Corfu, Gibraltar, etc. She exchanged names with the composite screw sloop *Lark* (ex *Kingfisher*) (see page 63) in May 1893. She has an armed Marine sentry on her forecastle and her hammocks are stowed along her bulwarks. The height of her bulwarks can be judged from the sailor standing by the midships accommodation ladder.

Opposite above

THE second picture is a view looking aft from the upper deck, and here the hammocks can be seen more clearly. The compass and chart table stand exposed on a decking level with the top of the bulwarks, whilst the quartermaster would man the wheel under the decking. The wheel is engraved with the words 'Hearts of Oak'. Knowing which rope was which among the complex of belaying pins, blocks and cleats must have been a daunting challenge to the young men when they first arrived onboard. She was sold at Malta in 1912.

Opposite below

HMS RAMBLER

A COMPOSITE screw gun vessel of the *Algerine* Class, the *Rambler* was launched by Elder, Glasgow, on 26 January 1880, and was completed as a survey vessel in 1884. Her Commanding Officer was Commander Hon Foley C P Vereker, who had previously commanded the survey vessel *Magpie* (see page 30). Of 835 tons, she differs from the *Magpie* in having sturdier lines and a single propeller, being better fitted for ocean, rather than coastal, operations. She has a square, workmanlike, bow with a large embrasure for a gun. In November 1884, she commissioned at Chatham and took passage to the Far East, and this photograph was probably taken as she called in at Malta. She was at Hong Kong in January 1886 and operated in the China Sea for several years. She was sold in January, 1907.

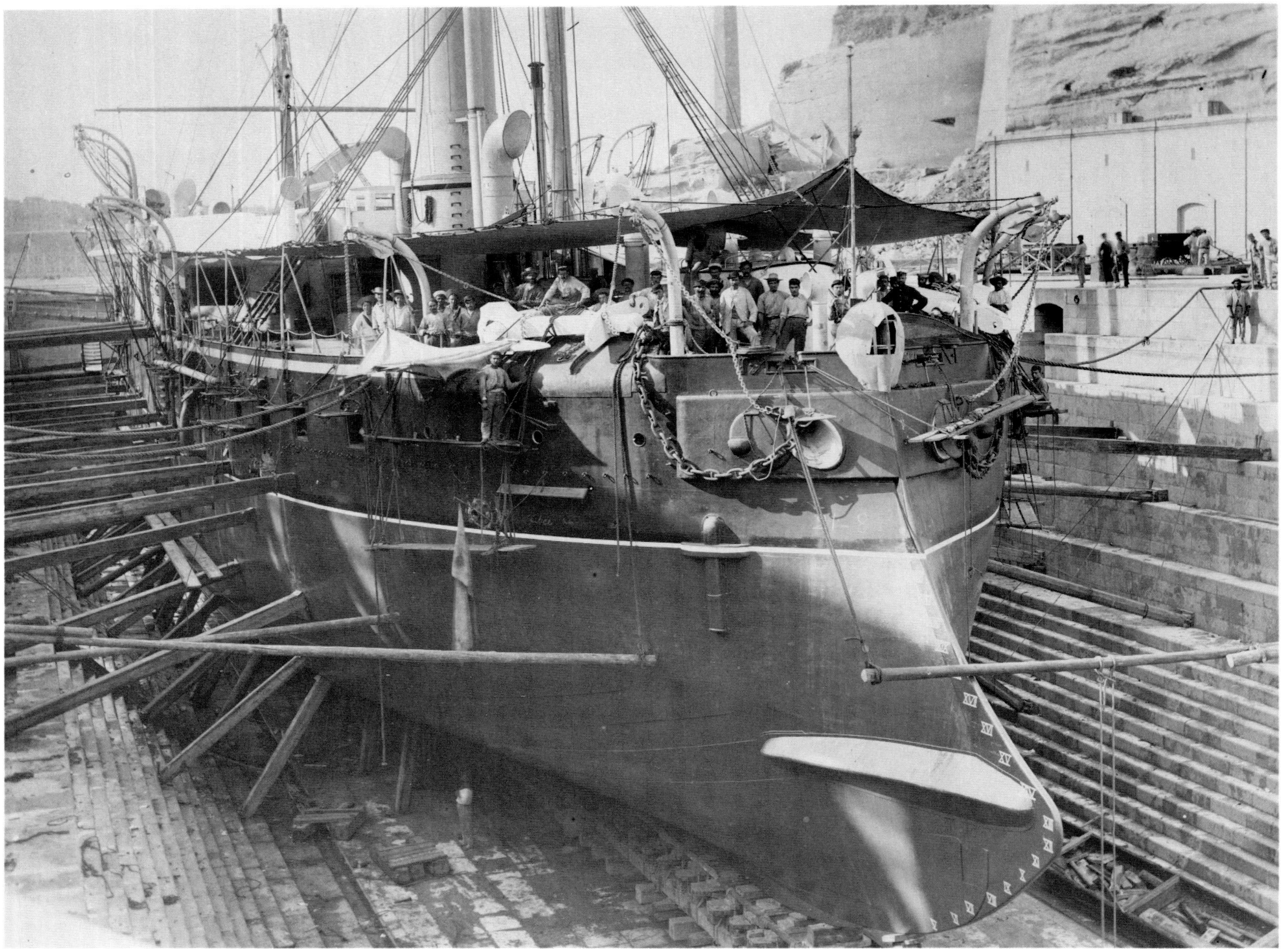

HMS ORION

1885-1886

AN ARMOURED corvette, the *Orion* was originally the Turkish *Bourdjou-Zaffer*, launched on 23 July 1879 by Samuda, Poplar, having been purchased on 13 February 1878 by the Admiralty during the Russian war scare. With four 12 inch muzzle loading guns, she, and her sister the *Belleisle*, were the only British ships to carry 12 inch guns in a broadside, and to have the entire armament in an upper deck central battery. Her armament was a higher proportion of the ship's displacement than any other British ship. She commissioned on 24 June 1882 for service in the Mediterranean and expected to sail with the squadron that bombarded Alexandria, but arrived two weeks too late. She landed parties to occupy Ismailia and her crew formed part of the naval brigade at Tel-el-Kebir. Having a low endurance she was of little use with the fleet, and was placed in reserve at Malta in May 83. In April 1885 she was commissioned and is seen here docking prior to service at Alexandria as the Senior Officer's ship. Her tall funnel replaced an original shorter one before she was completed. In the dock views her bow ram of eight feet of solid iron forging can be seen. This gave a huge bow wave and made the forecastle wet even in a calm sea. She had twin screws for manoeuvrability in shallow water and her after quadruple light machine guns can be seen on her stern. Her 'man overboard' lifebuoys with copper floats are suspended on her side aft. After docking she is seen in the white colour of the Mediterranean fleet at that time (1886). After a spell at Singapore, she returned to Malta for another period in reserve. In 1902 she became the parent ship for torpedo boats at Malta and in July 1904 she was the depot for the Malta RNR. Renamed *Orontes* as a store ship at Devonport in 1906, she was sold in June 1913.

HMS AGAMEMNON

A BATTLESHIP of 8510 tons, she carried four 12.5 inch muzzle loading guns and two six inch guns, one forward and the other aft (which can be seen clearly pointing over the stern). She was launched at Chatham on 17 September 1879 and her twin 18 ft diameter propellers were designed to drive her at 13 knots. After a period as a drill ship she commissioned at Devonport on 16 September 1884 for service on the China Station under Captain Samuel Long. She shadowed the Russian Cruiser *Vladimir Monomakh* on the way out, entering Yokohama in 1885 with the Russian cruiser's guns pointing at her waterline and not knowing whether war had been declared. She left Singapore in February 1886 and arrived at Malta on 17 July 1886. She had run aground several times going through the Suez Canal and she is seen here docking at Malta for the fitting of an extension to her stern and a stub to her rudder to improve her steering. She was known as the most unhandy ship ever to fly the White Ensign. In 1889 she was on temporary duty in the East Indies and took part in the blockade at Zanzibar. She returned to the Mediterranean, and in the broadside view photograph taken in May 1890, the extension put on the stern to aid her steering can be seen under her after boat. She also is carrying a 60 ft launch for torpedo attacks on her after superstructure. Note the way the anchor is stowed to allow clear arcs of fire for the guns, and the searchlights mounted over the bow. She served in the Mediterranean until 1892 and then paid off. She was sold in January 1903 and was broken up in Germany. She was the last battleship to have muzzle loading guns and the first to have a secondary battery as well as anti-torpedo guns (Nordenfelts and MGs).

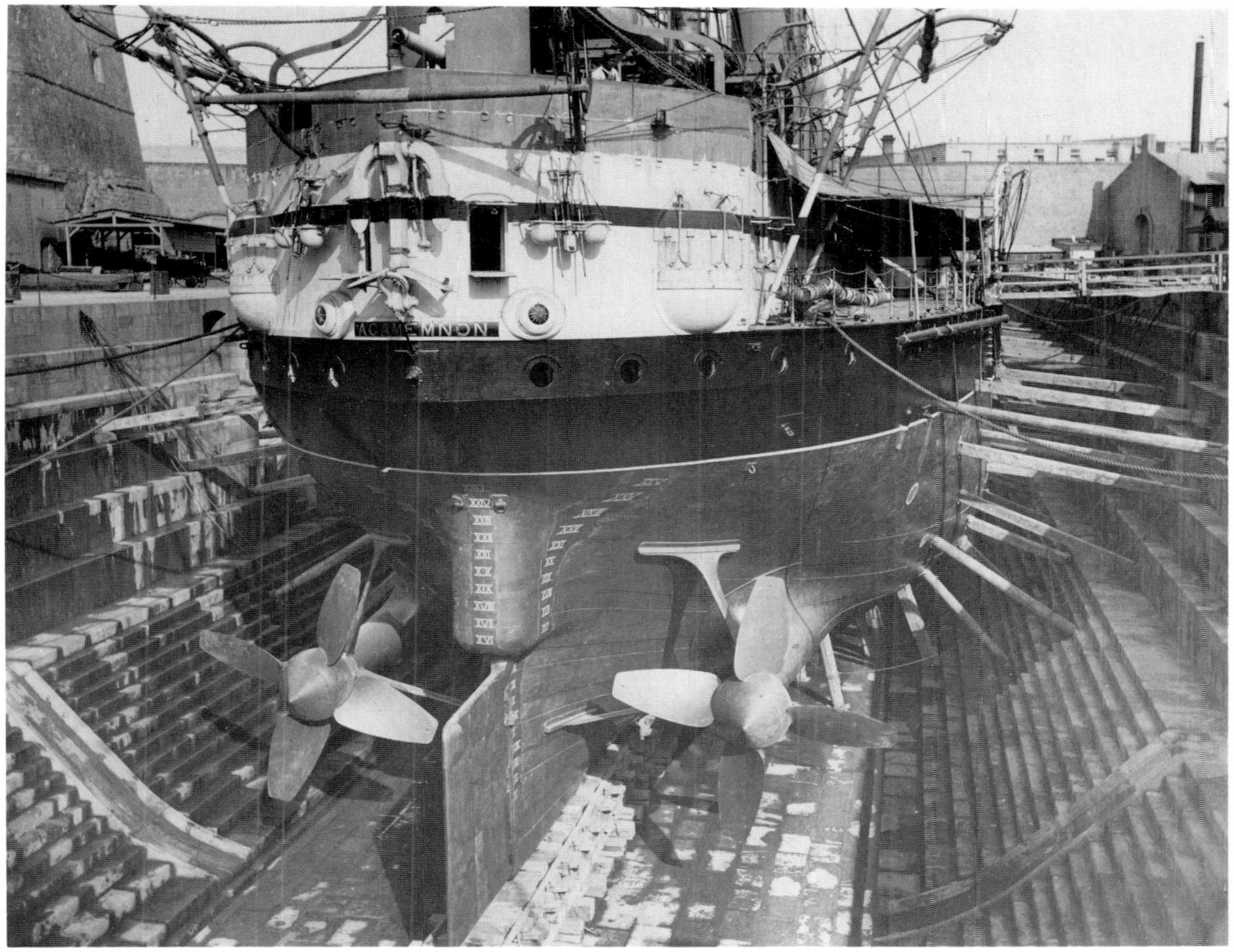
AGAMEMNON

HMS CARYSFORT

1886

A SCREW corvette of 2380 tons, the *Carysfort* was launched by Elder, Glasgow, on 26 September 1878. She was a slow steamer (13 knots) at a time when the *Iris* (1877) and *Mercury* (1879) could reach 18 knots. However, she could sail well, and with her sisters of the *Comus* Class she started the tradition of naming a class of ships with the same initial letter. This class also introduced metal hulls for smaller ships of the Navy. In 1880 she was part of Rear Admiral the Earl of Clanwilliam's Flying Squadron which sailed to Montevideo, Capetown, Australia, Fiji, Japan, China and back via Simonstown. Serving in the Squadron were two sons of the Prince of Wales, the Dukes of Clarence and York. As the Squadron headed home, the *Carysfort* detached at Gibraltar and headed East to join the Mediterranean Fleet for the Egyptian war. She landed men for the naval brigade at Sudan in 1884 and 1885. In August 1886 she relieved the *Orion* as the Senior Officer's ship at Alexandria. The *Orion* returned in the November, and *Carysfort* arrived at Malta in the December. She was sold to King, Garston on 15 December 1899.

TORPEDO BOAT 21

1887

A FIRST Class Torpedo Boat, TB 21, and her sister TB 22 (see page 119), was built by Thornycroft in 1885 to a slightly modified design of the *Childers*, which Thornycroft had built for Victoria, Australia. Both boats were sent to Malta on completion and remained there throughout their careers, TB 21 being sold in 1907. Fitted with three torpedo tubes and capable of 20 knots, she had to be reboilered with water tube boilers, and towards the end of her career had no deck tubes and was only capable of 15 knots. Her bow tube torpedo hatch, which created spray over the bridge, twin funnels and exposed upper deck can be seen here quite clearly as she lies alongside by some coal lighters. Her movements were not confined to Maltese coastal waters, for in the October of this year she and her sister visited Corfu escorted by the sloop *Gannet*.

Overleaf

HMS POLYTHEMUS

1886

A TORPEDO ram, launched at Chatham 15 June 1881, the *Polythemus* was the inspiration for H G Wells' *Thunderchild* in 'The War of the Worlds'. She was intended to escort a squadron of battleships in a protective role and was the only one of her type built. In 1885 she was in Admiral Hornby's Particular Service Squadron which exercised torpedo attacks and defences against them. During these exercises the *Polythemus* charged the boom at Berehaven, passing right over it without any damage to herself. Her captain from 1882 was Commander Edwin J P Gallwey (later to command the *Hecla*—see page 50). In April 1886 she left Devonport and sailed to Malta from Gibraltar escorting torpedo boats No 21 and 22 (see above and page 119) From Malta she went on to join the Fleet at Suda Bay. She was fitted with a torpedo tube in her ram which can be seen here. She was the first seagoing warship with a submerged, sideways firing torpedo tube. Indeed, her main armament was her five 14 inch torpedo tubes, the ram being a spectacular extra. With her armoured, curved deck, speed of 18 knots and many reloads for her torpedo tubes, she presented a real threat by virtue of her weaponry rather than by physical contact. She had bow rudders, which can be seen under her ram bow. She was known as 'One Eye' by the sailors. She was sold in July 1903 to Cohen.

XXI
XX
XIX
XVIII
XVII
XV
XIV
XIII

OSBORNE

OSBORNE was one of the Royal Yachts, being launched at Pembroke Dock in 1870. Her paddle propulsion was the choice of Queen Victoria, who preferred its rhythmic beat. *Osborne* was, however, the last of the Royal paddle yachts. After Prince Albert has died, the Queen confined her voyaging to the coastal waters around the United Kingdom, leaving the travelling to distant waters to the younger members of the Royal Family. As can be seen in this photograph, the boats bear the badge of the Prince of Wales. Her large wheel can be seen mounted abaft the mizzen mast. In 1877 her captain reported seeing a large, unidentified sea monster off the North Coast of Sicily. Her Captain at the time of this photograph (about 1877) was the Hon Assheton G Curzon-Howe, who later returned to the Mediterranean Fleet in command of the battleship *Revenge* (see pages 74). The *Osborne* was sold in July 1908, and was broken up at Felixstowe.

HMS DREADNOUGHT

2 January 1888

SEEN here cleared for action, with her guardrails and forward bulwarks lowered and the light guns on the 'flying deck' manned, the *Dreadnought* shows herself to be the last true ironclad. She had recommissioned at Malta on 13 December, her Commander (Executive Officer) being HRH Prince Louis of Battenburg, and another officer serving onboard was HRH Prince George of Wales. Originally to have been the third ship of the *Devastation* Class and named *Fury*, the design was altered and the *Fury* cancelled and the *Dreadnought* built in lieu. One major improvement was the introduction of artificial ventilation, and she was also better subdivided. The armour, up to 14 inches thick on the sides, turrets and conning tower, comprised nearly 34 per cent of her displacement of 10,820 tons. 1500 tons heavier than her half sisters, and half a knot faster, she was very wet at sea, but proved popular through out her career even though she was not re-armed to bring her up to date. Having been launched at Pembroke Dock on 8th March 1875, she completed in February 1879, commissioned for trials and then was placed in reserve until 1884. In the October of that year she commissioned for the Mediterranean where she spent the next ten years. After a refit in 1894 at Chatham, she spent some time on Coast Guard duty and commissioned for manoeuvres and in 1902 acted as a depot ship for torpedo boats. She was sold in 1908 and broken up at Barrow and Preston. The troopship lying beyond her is probably the *Serapis* (1866), sister to the *Malabar* (see page 51).

HMS MELITA

1888

A COMPOSITE sloop of the *Mariner* Class, *Melita* was laid down at Malta Dockyard on 18 July 1883. The decision to build her there was taken to give employment on the island, although her construction cost about a fifth more than her sisters built in the Royal Dockyards at home. She was launched on 20 March 1888, by Princess Melita, daughter of the Duke of Edinburgh, who was the Commander-in-Chief, Mediterranean at that time. The *Melita* was converted to a Boom Defence Vessel in 1905, became a Salvage Vessel (renamed *Ringdove*) in 1915 and was sold to the Falmouth Docks Board in 1920.

HMS AUSTRALIA

April 1890

AN ARMOURED cruiser of the *Orlando* Class, the *Australia* was completed in October 1888, having been built by Fairfield. She carried a single 9.2 inch gun on the forecastle and another on her quarterdeck, and had five 6 inch guns on each side. Designed for 18 knots from her triple expansion engines and twin screws, she achieved 18.8 knots on trials. The class was built for trade protection rather than for operating with the Fleet, a task for which they were too slow. Originally there were problems with the supply of their main armament, and some of the class appeared at the Jubilee Review with wooden guns. At first there were no shields to the 9.2 inch guns, but by the time of this photograph they had been fitted. The Paddle Tug alongside appears to be the *Hellespont* (137 tons) (purchased 1878) one of the three paddle tugs at Malta at the time. *Australia* commissioned at Chatham in November 1889, and then sailed for the Mediterranean where she stayed till 1893. For the ten years following she was on Coast Guard service in Southampton Water. She was sold in 1905.

HMS CAMPERDOWN

April 1890 July 1893

A BATTLESHIP of the *Admiral* Class, she was a barbette ship and was launched at Portsmouth Dockyard in 1885. Completed late in 1889, the delay being caused by the slow production of her 13.5 inch guns, she proceeded to the Mediterranean in that year, where she became the temporary Flagship of that Fleet until relieved by the *Victoria* on 24 April. She is seen above wearing the flag of Vice Admiral Sir Anthony H Hoskins KCB. She sailed for the Channel on 3 May to become the Flagship of the Channel Squadron. In 1893 the *Camperdown* had returned to the Mediterranean and Rear Admiral Hastings Markham transferred his flag to her from the *Trafalgar*. During manoeuvres off Tripoli in 22 June she collided with the *Victoria* (see pages 52 and 53). The *Victoria* sank. The external damage to her bows can be seen on the right, and the effectiveness of the ram bow was well demonstrated by her effect on *Victoria*, which was penetrated to about 6 feet about 12 feet below the water line, the hole being about 100 square feet. *Camperdown* was herself saved by the quick action of her carpenter who built a temporary bulkhead to stop water pouring into her. As it was, her metacentric height was reduced to virtually nil, and she was fortunate that it was a calm day. Repairs took until September. By 1908 she lay at Harwich as a berthing ship for submarines, and she was sold in 1911.

HMS HECLA

April 1890

LAID down as the *British Crown*, she was bought by the Admiralty and launched as the *Hecla* in 1878 by Harland and Wolff, Belfast. She carried 6 second class torpedo boats, which can be seen here hoisted inboard with their funnels lowered. She also serviced other torpedo boats and attended to the torpedo and mining requirements of the Fleet. Commanded by energetic captains, she did much to develop torpedo tactics in the Navy. She had come out to the Mediterranean in November 1889, the usual pattern of her operations being to exercise off Berehaven in the summer, and off Malta in the winter. She is seen here arriving at Malta on 15 April from Platea. During the Egyptian War of 1882-4 her Captain, Arthur Knyvet Wilson, was awarded the VC. The first VC had been awarded to Mate Lucas of the previous *Hecla* in 1854. In 1885 the *Hecla* was part of Admiral Hornby's Baltic Squadron which tested the new torpedo boats and anti-torpedo equipment off Portland. In 1908 she was refitted as a destroyer depot ship, in which capacity she served until she was broken up at Preston in 1926.

HM Indian Troopship MALABAR

22 November 1890

THE *Malabar* was one of five troopships built in 1866, the Indian Government meeting the bill, hence these ships carried the Star of India in gold on each bow. All were painted white with a yellow funnel, and they could be distinguished by a different coloured band around the hull. *Malabar* had a black band, whilst *Jumna* had red, *Crocodile* yellow, *Serapis* (see page 45) green and *Euphrates* had blue, which later became the traditional colour for troopships. Each ship could carry 1200 people, being the number in an infantry battalion when accompanied by the families of the married men. Before the Suez Canal opened, in 1869, two of these vessels would carry troops from England to Alexandria. The troops would then travel by rail (the railway line opened in 1858) to Suez, where they would embark in *Malabar* or one of her other two sisters for the journey to Bombay. The sails would steady the ship and also add the extra knot for the voyage. She is seen here with her upper deck crowded with people catching their last glimpse of Malta in the evening autumn sunshine as she heads for Portsmouth. The round trip from Bombay to Portsmouth and back to Bombay took her 70 days. She is fast approaching the end of her trooping days for by 1894 only she of the of the five sisters was in reasonable condition. In 1896 she was sold to the Admiralty and in 1897 she became a receiving ship at Bermuda. In May 1901, she replaced the old *Terror* there as the Base Ship. In 1914 she was placed on the sales list, but was not sold until January 1918, at Bermuda.

HMS VICTORIA

COMPLETED in 1888 at Elswick by Armstrongs, but delays with her large guns delayed her first commissioning, which was at Chatham on 19 March 1890. She sailed from Portsmouth on 15 April to become the Flagship of the Mediterranean Fleet when she arrived on station on 24 April. Admiral Hoskins transferred from the *Camperdown* (see page 48), his flag can be seen flying from her topmast in that same month. She was the first battleship to have triple expansion engines, and was the first built by Armstrongs. Her two 16.25 inch guns can be seen forward, whilst she has a single 10 inch gun aft. Her low forecastle made her very wet forward. Three months later her funnels were raised by 17ft to improve the natural draught to the boilers.

In January 1892, she ran aground at Platea and was floated off 6 days later, after over 1200 tons of equipment had been removed, and with the assistance of the *Edinburgh* and *Dreadnought*. Sailors can be seen scrubbing down her sides as she docks down for a refit after this event in the Hamilton Dock, she being the first ship to enter it. The next photograph, taken in May, shows the ship spruced up and ready for sea again. Her twin funnels with their extra height, signal masts and semaphore arms can be seen here.

She recommissioned on 1st April 1893, and was leading the Fleet from Beirut to Tripoli as the Flagship of Vice Admiral Sir G Tryon on 22 June when she was rammed by the *Camperdown* during manoeuvres. Not all the watertight doors were closed, and before they could be shut water was washing in through the gunports. Admiral Tryon, 22 officers and 336 men were lost. Among the 294 saved was the ship's Commander, John Jellicoe.

HMS FEARLESS

12 February 1891

LAUNCHED by the Naval Construction and Armaments Co at Barrow on 20 March 1886, and completed in the July of the next year, *Fearless* was a Torpedo Cruiser of 1580 tons. She, and her sister ship the *Scout* (see page 100), were planned to operate with the fleet, but they proved too slow at their 17 knots, and were also poor seaboats. As a result, they were relegated to commerce protection. She had a fixed torpedo tube in the bow, and traversing torpedo carriages under the forecastle and poop. The ports for the traversing carriages can be seen in her side, just ahead of the forward 5 inch gun sponson, and abaft the after 5 inch gun sponson. She spent from 1888 to 1900 in the Mediterranean, except for a period in 1899 when she was temporarily detached to the Cape of Good Hope. She is seen here shortly before she set out on a cruise to Volo, Smyrna and Salonika. She was refitted in 1900, then spent a while at Sheerness before sailing for the China Station where she remained until 1904. She was laid up in 1905 and was sold in the July of that year at Portsmouth.

Previous pages

GRAND HARBOUR

IN THE foreground lies the Flagship of the Mediterranean Fleet, the *Admiral* Class battleship *Camperdown* (see page 48), while beyond her, to the left, is her half sister, the *Collingwood*. The *Collingwood* was the forerunner of the *Admiral* Class, and was the first British battleship to have breech loading guns in barbettes. Her 12 inch guns were fitted 22 ft above the waterline, which gave an advantage when working them in a seaway. However, her underwater shape was such that the seas washed over her, hitting the barbette and spraying the guns as well as flooding the messdecks. The separation of her two main barbettes and grouping together into batteries of the secondary and tertiary armament became the pattern for future construction. Launched at Pembroke in 1882, she was completed in 1887 and commissioned for the Jubilee Review. She then paid off, but took part in manoeuvres in 1888 and 1889. In the November of 1889 she sailed for the Mediterranean where she spent eight years before Coast Guard service and reserve. She was sold in 1909. The third ship in the photograph is the *Edinburgh*. She was a *Colossus* Class turret ship, these being the first British battleships built of steel, and to have compound armour generally rather than iron. This class also introduced breech loading for their main armament of four 12 inch guns. She was also launched at Pembroke in 1882, and was completed in 1887, commissioned for the Jubilee Review and a month later sailed for the Mediterranean, where she stayed until 1894. She then carried out Coast Guard duties and spent time in reserve and for a period was a gunnery tender at Sheerness. In 1908 she was used for target firings after being fitted with modern armour, and this led to Jellicoe ordering a redesign of armour piercing shells. Regrettably this redesign was not carried out. She was also sold in 1909.

PLASSY (Royal Indian Marine)

1892

ONE of two *Sharpshooter* Class Torpedo Gunboats built for the Royal Indian Marine, which replaced the Royal Indian Navy in April 1863. Armstrongs built both the Indian vessels and two for Australia, whilst all the Royal Naval vessels of the class were built in the Royal Dockyards. Of 735 tons, they carried two 4.7 inch guns and three torpedo tubes. Their two shafts gave them 19 knots. Built without armour their machinery was protected from machine gun fire by bunkers of coal placed on either side of and above the boiler rooms. Some of the vessels of this class were fitted with different makes of boilers to provide a comparative trial and as a result Belleville boilers were chosen for the *Canopus* Class battleships and the cruisers *Powerful* and *Terrible*. The *Plassy* had been launched in the previous November, and completed in February 1892, and is seen here on her passage out to East Indies. She was sold in 1904.

HMS ARETHUSA

13 June 1893

A *LEANDER* Class cruiser, built by Napier at Glasgow, she was launched on 23 December 1882 and completed in 1886. She was held at Chatham until May 1893, when she commissioned for the Mediterranean. She left Portsmouth on 30 May and reached Malta on 9 June. She was due to sail to join the Fleet on 23 June, but remained at Malta due to the collision between *Victoria* and *Camperdown*. Off Malta she carried out experiments with the new explosive, lyddite. Armed with ten 6 inch guns, the forward and after guns on each side were mounted in sponsons to improve the arcs of fire ahead and astern. Originally planned as despatch vessels, the class were redesignated before completion. Their barque rig was more of a problem than it was worth, and it was reduced to barquetine (as seen here), and then to schooner before it was finally removed altogether. She can be seen with windsails rigged to cool the messdecks, and with the church pennant hoisted indicating that a church service was being held onboard. Hoisted on her port side is Torpedo Boat 84, built in 1884 by Yarrow. After leaving the Mediterranean she spent two years at Chatham before going out to the Pacific, where she served until 1903. She was sold in April 1905.

HMS SKIPJACK

April 1894

A *SHARPSHOOTER* Class Torpedo Gunboat, the *Skipjack* is seen here alongside in Dockyard Creek. She was launched at Chatham in April 1889, and completed two years later. Her 4.7 inch guns are fitted forward and aft under the awnings. This gun was selected for the class, rather than the designed 4 inch gun, as trials showed it could be fired five times faster. One of her five torpedo tubes was fitted in the bow. The port pair can be seen abaft the after funnel, whilst the other pair are to starboard and further forward. This class suffered from hull weaknesses at first. They were not fast enough (at 19 knots) to work with the fleet as planned, but they were useful ships. *Skipjack* had commissioned at Chatham for the Mediterranean in January 1894. She had been with the *Aeolis* and *Gleaner* to Argostoli, and after a maintenance period at Malta in April, she sailed as part of the Cruiser Division to Corfu, rejoining the main fleet in June. She remained in the Mediterranean until 1897. She was one of several of this class converted to minesweeping duties in 1908-9, and she was sold in April 1920.

PICKING UP SURVIVORS

22 June, 1893

JUST ten minutes after the *Victoria* had been rammed by the *Camperdown* (see pages 48 and 49), she sank. In that time her crew had mustered four deep on the port side of the upper deck, though the engine room and boiler room crews had had to stay at their posts, and the teams trying to combat the inrush of water were working waist deep. *Camperdown* had a struggle to remain afloat, but fortunately it was a calm day otherwise the other ships of the Squadron would have been even harder pressed. On the left of this scene is the *Inflexible* (see page 20). The Commander of the *Inflexible* at this time was Commander G H Cherry, who became immortalised by the award of the Cherry Medal to Officers who served under him in the *Argonaut* in 1900-4. The *Barham* (see page 96) is nearer the camera, and then the *Edinburgh* (see page 54) with the *Fearless* (see page 56) on the right. The *Barham* and *Fearless* had not been in the two lines of ships manoeuvring, but there had been a total of 13 ships in the Squadron. It is interesting to compare the loss of life (359) from one ship with the proportion of ship's companies lost in the less benign conditions of weather and stress in the Falklands War ninety years later. Clearly the efforts made in improving lifesaving techniques, including the ability to swim, have been worthwhile.

HMS EDGAR

2 July 1893 (with survivors of Victoria*)*

NAMESHIP of a class of First Class Cruisers, *Edgar* was launched at Devonport on 24 November 1890, and completed in March 1893. She had just arrived on the Mediterranean Station when she was involved in the manoeuvres in which the *Victoria* was lost. She is seen here returning to Malta with survivors from the *Victoria*. Her single 9.2 inch gun on the forecastle and single 6 inch guns mounted along her sides can be seen plainly in this view. The fitting over her bow is a boom for the torpedo nets. In 1894 she went out to the China Station for 3 years. After that she was employed transporting crews between England and China and later she was a training ship for boys. During the First World war she was fitted with bulges against torpedoes and with net sweeping gallows. Her 9.2 inch guns were removed for fitting to the M15 Class monitors. She supported the army at the Dardanelles and was at the evacuation of the Gallipoli Peninsula. She was finally sold in 1921 and broken up in 1923. The machinery of this class proved exceptionally reliable.

HMS HAWKE

November 1893

THESE two photographs of the *Hawke* in dry dock show clearly the fittings for supporting her anti-torpedo nets which the Navy had developed a few years before. Although the nets slowed the ships because of their drag in the water, and net cutters had been introduced on torpedoes, thus rendering the nets ineffective, the nets were retained for a long period, possibly because of their value in providing convenient labour intensive evolutions. An *Edgar* Class cruiser, the *Hawke* had been launched at Chatham in 1891 and had commissioned there on 16 May 1893, under the command of Captain Pelham Aldrich. She arrived at Malta on 13 July carrying Admiral Sir M Culme-Seymour, who then assumed command of the Fleet. After a cruise to Greece and Italy, the *Hawke* docked prior to sailing on 25 November to Alexandria to relieve the *Arethusa* as Senior Officer Egypt. A gun limber is on her quarterdeck. She remained in the Mediterranean until 1899. Having reliable machinery, this class saw an unusual amount of sea service for ships of their size. After a collision with the White Star liner *Olympic* in the Channel on 20 September 1911, *Hawke* was given a straight bow. She was torpedoed by Weddigan in *U9* on 15 October 1914, with the loss of 524 men, less than a month after Weddigan had sunk the *Hogue*, *Aboukir* (see page 122) and *Cressy*.

Overleaf

HMS CRUISER

LAUNCHED as the *Kingfisher* at Sheerness on 16 December 1879, she was a composite screw sloop of 1130 tons. Her men were in the Naval Brigade that captured Witu in October 1890. In 1892 she was renamed *Lark*, and on 18 May 1893 she commissioned at Malta as the *Cruiser*. The wood screw sloop *Cruizer* was renamed *Lark* at the same time (see page 34). Both were employed as training ships for ordinary seamen. In these pictures she gives a fine display of smartness, as befits her training role. The two views of the upper deck show the high bulwarks, and the polished guns, fancy ropework, and the brass bound buckets spelling out the ship's name. In the gunroom, there are midshipmen's chests, and the rum barrel and the barricoe for bringing up the rum from the store. The clock indicates there are 25 minutes to 'Up Spirits'. The fourth view is of her tacking her way out of Grand Harbour under full sail whilst a harbour tug passes inward bound, taken on a blustery May day in 1897, as she was departing on a cruise to Corfu. She was sold in 1919.

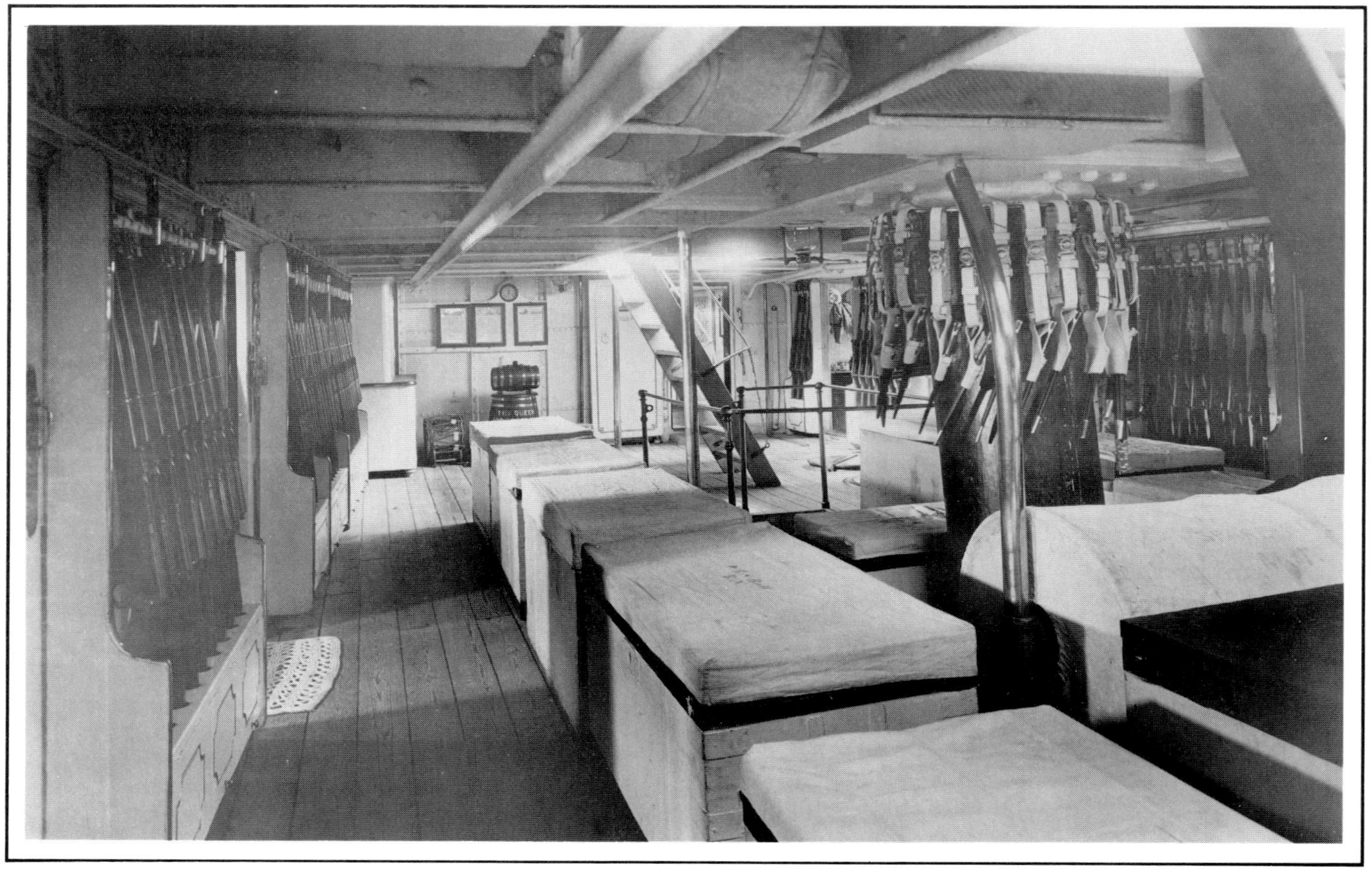

HMS BULLFROG

October 1894

A COMPOSITE Screw Gunboat of the *Banterer* Class, she displaced 465 tons and was only 125 ft long. Her reciprocating engine drove her single shaft at 9½ knots with its 440 indicated horse power. This class were distinguished by their vertical stems, which can be seen here. She had been launched at Pembroke Dock in February 1881. She has 20 pounder breech loading guns as bow chasers, and the hatch for the port gun can be seen forward. She also had two 64 pounders on the centreline. She has a searchlight fitted abaft the funnel and her binnacle is forward of the mizzen mast. The *Bullfrog* and *Firefly* (screw gunboat) sailed from Devonport on 4th October 1894 and arrived at Malta on 22nd October. She was relegated to harbour service there in 1895 and was renamed *Egmont* in March, 1923—relieving the *Firefly* as the 'name ship' for the Base at Malta. She was renamed *St Angelo* in July 1933 and was sold in that year, 52 years after launching.

Previous pages

HMS ARDENT

1896

ARDENT was a 'Thornycroft 27 knotter' destroyer, launched in October, 1894, and completed at Chiswick in April 1895. She had twin screws and reached 27.97 knots (mean speed) on trials. Her 12 pounder, forward, and five 6 pounders can be seen here. Her two torpedo tubes on the deck are hidden by her crew. She spent 16 months in the Mediterranean, after which experience it was reported that life onboard was hardly comfortable, but it was endurable, and only one man had to be invalided home in spite of the extremes of heat and cold to which the crews had been subjected. She proved an excellent seaboat and in moderate weather could go anywhere, though in a head sea and when hard pressed it was hard to prevent water going down below. She is also reported to have steered very well both ahead and astern. At this time she was commander by Commander G H B Mundy, instead of by a Lieutenant, and she was a tender to the *Ramillies*. In the September she was joined by *Banshee*, *Dragon* and *Bruizer* (see page 107). In 1904 she was one of the force shadowing the Russian Baltic Fleet on its way to the Far East for the Russo-Japanese war. In 1906 she collided with Torpedo Boat 80 and sank her. She was sold in October 1911.

HMS TRAFALGAR

March 1896

IN this scene the after 13.5 inch guns of the *Trafalgar* are being mounted. The guns were 36 ft long, weighed 67 tons and fired a 1250 lb projectile. With a trained crew each gun could fire once every two minutes. Having been launched at Portsmouth in September 1887, her completion was delayed until March 1890, awaiting her main armament. She sailed that year for the Mediterranean, where she spent most of her active life, mainly because the seas there were more suitable for her low freeboard. Her funnels were raised in 1891, and were fully cased, unlike her sister, the *Nile* (see page 76). Captain R C Prothero (known as 'Prothero the Bad''—see text for page 76) took command of her in April 1896. She was Flagship of the Rear Admiral, Mediterranean Fleet, but the flag had been shifted to *Barfleur* from March to August whilst she was refitting. She paid off at Portsmouth 18 months after this photograph was taken, and spent the next few years as port guardship and then was placed in reserve. She became the turret and submerged torpedo tube drill ship at Sheerness in 1907, and, after being relieved by the *Vengeance* in 1909, was sold in 1911.

HMS HOWE—Sailing Model

March 1896

IN this photograph the Midshipmen of the *Howe* are being instructed in rigging. The instructional aid is a ship's boat made up with the rigging of a full rigged ship, even down to the detail of the figurehead. Considerable effort must have gone into the preparation of this model, with its ensign and commissioning pennant. Full marks must be given to her Chief Boatswain, J C Andrews, and her Boatswain, Benjamin Flevill. The ship's engine room telegraph can be seen above the canvas screen. At this time the *Howe*, an *Admiral* Class battleship, was within nine months of returning home to pay off. She had been launched at Pembroke Dock in April 1885, and completed in July 1889. Whilst with the Channel Fleet on a Mediterranean cruise in 1892 she grounded on a rock off Ferrol. She spent four months stranded there, but was able to steam herself home after she had been salvaged. Despite the period aground, repairs took only 4 months, which must be a tribute to her construction and her builders. She joined the Mediterranean Fleet in 1893. After returning home in December 1896, she was port guard ship for a period and then placed in reserve and only brought forward for manoeuvres. She was sold in 1910.

Torpedo Boat No 56

1896

BUILT by Thornycroft at Chiswick in 1886, *TB* number 56 was of 60 tons, and was 127½ feet long. Torpedo boats had been ordered in quantity at that time because of a fear of war with Russia, and efforts were made to balance the building programmes of the French and Russians. Originally to have had only one fixed bow torpedo tube and a mainly gun armament for use against Russian torpedo boats, construction was delayed while the bow was modified and two pairs of deck mounted tubes were fitted instead of 3 pounder guns. *TB* 56 capsized on the night 17/18 May 1906, whilst being towed by the cruiser *Arrogant* off Damietta (31° 46′N - 031° 32′E). The crew seen here are wearing *Hibernia* cap tallies, the Base Ship at Malta at the time, whilst beyond can be seen an *Astraea* Class cruiser and the battleship *Barfleur* (1892).

HMS ANSON

3 June 1896

THE *Anson*, an *Admiral* Class battleship, is seen entering Malta from Limassol with her paying off pennant flying. She paid off at Malta and recommissioned there under the same Captain (W F S Mann AdC)—refitting until September. Preparations are going ahead for her entry and a sailor can be seen rigging the lower boom under the bridge. The fourth and last of her class, she was launched at Pembroke Dock in February 1886, but completed second, in May 1889 despite having waited two years for delivery of her guns. In this view her secondary battery of six inch guns can be seen, as can her six pounders. She had her first three 6 pounders in the battery under the boats grouped together with the fourth spaced aft, which distinguished her from her sisters. She spent from 1893 to January 1900, in the Mediterranean, and then returned home to be in the reserve. She was sold in July 1909 and was broken up at Upnor.

THE FLEET IN HARBOUR

July 1896

IN this general view looking southwards towards The Marsa and Corradino Heights the Fleet lies quietly under the summer Sun with its awnings spread. The Flagship, lying at the head of the line, is the *Barfleur*. She was launched at Chatham in August 1892 and completed in June 1894. With her sister, the *Centurion*, she was a reversion to a smaller size battleship after the *Royal Sovereign* Class. Their 10 inch guns, which were the first high trajectory heavy guns in the Navy, had hooded barbettes with all round loading. Their secondary armament of 4.7 inch guns was criticised as being too light, and they were replaced by 6 inch guns when reconstruction took place in 1901. These two small battleships were built to counter Russian cruisers operating in Far Eastern waters and were not meant to take a place in a fleet action. Beyond her lies a *Royal Sovereign* Class battleship, and astern of her is the *Admiral* Class battleship *Howe* (see page 70) just returned from a period at Alexandretta. Further astern lies an *Edgar* Class cruiser, with a sister further inshore, (both the *Hawke* (see page 62) and the *Theseus* were in the Mediterranean at about this time) and a smaller cruiser of the *Astraea* Class (see page 80) is at the end of the line.

HMS REVENGE

August 1896

A *ROYAL SOVEREIGN* Class Battleship, the *Revenge* was undocked at Jarrow-on-Tyne by Palmers in 1892. She was completed in 1894, but remained in the Fleet Reserve until commissioned in January 1896, when she became the Flagship of Rear Admiral Alfred T Dale, who commanded the Particular Service Squadron formed at the time of European unrest following the Jameson raid in South Africa. Seen here whilst in that role, the Squadron having left Queenstown in May for Lisbon, joining the Mediterranean Fleet in June. After a spell of cruising independently, the Squadron rejoined the Mediterranean Fleet on 14 September. After recommissioning at Portsmouth she became the Second Flagship of the Mediterranean Fleet in the November. Her Battle Honours can be seen on the after screen, and the old style semaphore signalling arms can be seen on the after bridge wings. Her side by side funnel arrangement and 13.5 inch guns can also be seen in this view.

HMS REVENGE

July 1899

THE *Revenge* is seen here sailing for Girgenti, Sicily when Flagship of the Second in Command the Mediterranean, which she was from November 1896 to June 1900. Earlier in that commission she took part in Cretan Operations 1896-1897. Her secondary armament of six inch guns are clear in this view, as is the hatch for her after torpedo tube, situated just under her sternwalk. Rescued from the breakers at the outbreak of World War I, she was used for bombardment duties, her main armament's bore being reduced to 12 inches and the range being increased by heeling the ship. She was renamed *Redoubtable* in 1915 to release her name for a new battleship. She was broken up in 1919.

HMS NILE

THE last single citadel turret ships to be built, the *Nile*, and her sister ship the *Trafalgar*, were the heaviest British battleships built up to that time, and of their 11940 tons, one third was armour. Launched at Pembroke Dock in March 1888, *Nile* carried out trials in July 1890 without her guns and mountings. She was completed in 1891, and after manoeuvres sailed for the Mediterranean, where she remained until 1898. The photographs of her in dock in 1892 show her forward pair of 13.5 inch guns and low (12 ft) freeboard. In 1892 her Commander was R C Prothero (known as 'Prothero the Bad'), who didn't always see eye to eye with his Captain (Gerry Noel). It is said that the Captin sounded general quarters just as the Commander had the ship ready for painting, and half way through the drills the Captan declared himself killed, the Commander then sounded 'secure, painting parties fall in'. When steaming she would pile up water even in a calm sea and the decks would be awash. The funnels were extended before she completed, and the outer casings did not reach the top, which distinguished her from her sister whose funnels were fully closed. The man overboard lifebuoy aft has two tubes attached filled with calcium. When the buoy was dropped, the tops of the tubes were automatically opened and the sea water would cause the calcium to flare and smoke, helping the ship to find the buoy. A view of her at a buoy in Grand Harbour in October 1896 shows the deck layout, with sailors relaxing in the sun. She sailed to rejoin the Fleet at Salonica on 16 October. After returning home in 1898 she spent a period as Portguard ship at Devonport and then went into reserve. She was sold in July 1912 and broken up at Swansea and Briton Ferry.

HMS ASTRAEA

ASTRAEA was the name ship of a class of Second Class Protected Cruisers. She was built at Devonport Dockyard, being launched on 17 March 1893. She completed in November 1895, and proceded straight to the Mediterranean. In the broadside view, taken in August 1896, she is seen cleared for action and with her boats away. Her forward 6 inch gun can be seen plainly, but the after gun is obscured by the party of marines. In this view the secondary armament of 4.7 inch guns can be seen mounted on the same deck as the main armament, an improvement over previous classes where these guns were lower, and hence suffered from the weather. This class, were larger but with virtually the same armament and speed as the previous class (the *Apollos*), but were much better seaboats.

In the next view the Sick Bay is seen in 1897. Sick Bays had been introduced into the Navy 100 years before, in the Mediterranean, by Lord St Vincent. The term 'bay' derived from the shape of the ship's bow where the sick were tended, it being the airiest place when the ship was lying at anchor. The Sick Berth branch had only been formed on 17 October 1884. The anchor emblazoned counterpanes have only recently (1970s) gone out of service, having been replaced by more colourful, but less interesting, tri-service coverings.

In the last view her upper deck is seen with its mix of 4.7 inch guns and 6 pounders. The sailors' and marines' rig of the day can be seen here, and the marine is standing by a field gun. The battleship *Trafalgar* (see page 69) lies in the background.

The *Astraea* had been Senior Officer Syrian Coast in early 1897, and in the May was lent temporarily to the Cape of Good Hope Station. On 9th September she left Malta for Alexandria to be the Senior Officer Egypt. She remained active during the First World War, though most of her sisters had been reduced to secondary duties by then. In August 1914, she bombarded Dar-es-Salaam, rendering the port useless as a base for the *Konigsberg*. She finally paid off in 1919, was sold in 1920 and was broken up in Germany.

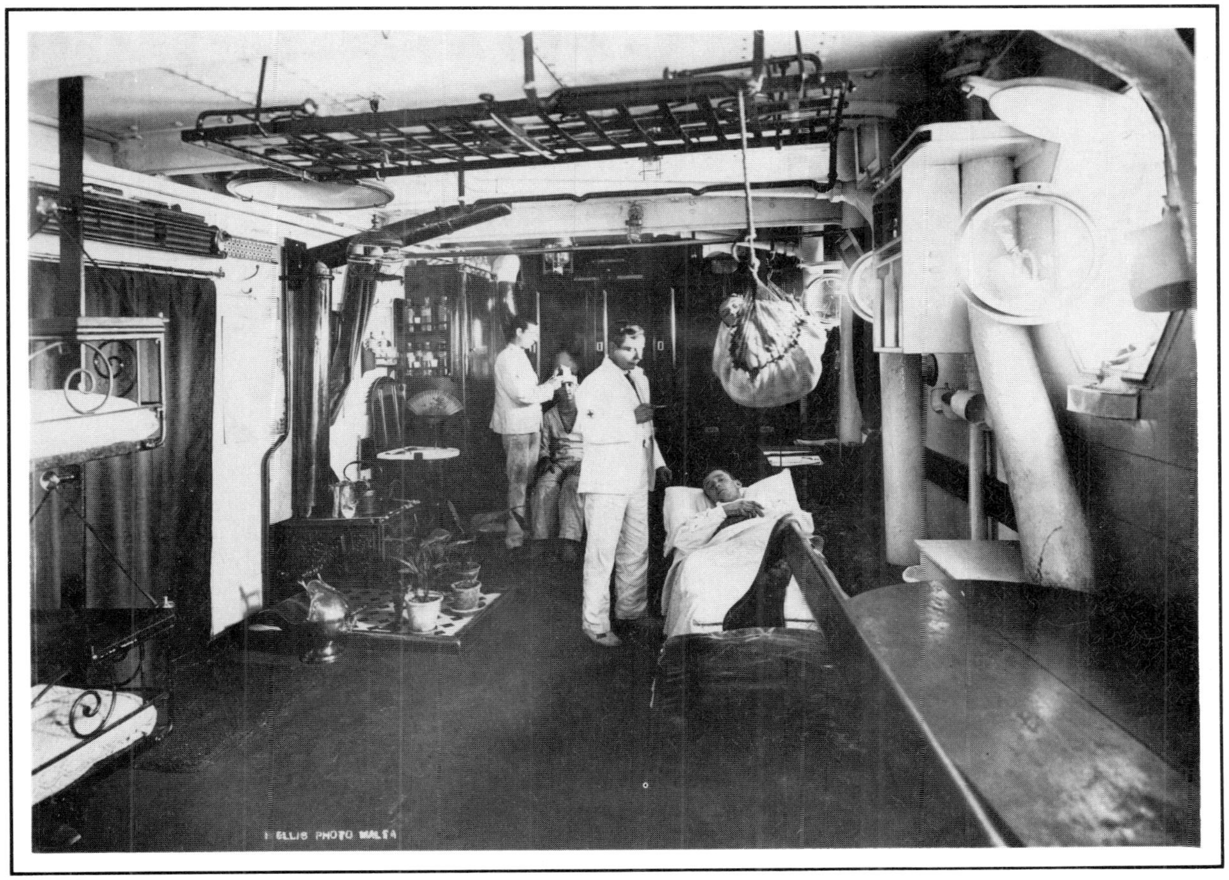
ELLIS PHOTO MALTA

HMS HANDY

October 1896

THIS 275 ton destroyer was built by Fairfield, being launched on 9th March 1895. Just 200 ft long, her engines developed 3800 horse power and she achieved 27.04 knots on trials. This must have been a challenge to the stokers of these small vessels, as the boiler furnaces glowed white hot and they had to wear dark glasses to operate them. She has shorter funnels than the *Ardent* (see page 66) , and her stern is squarer, with the rudder showing above the waterline. This was in the days prior to standardisation. Here she is seen when on passage to China with her sister ship, the *Hart*. They had been escorted to Malta by the cruiser *Blenheim*, arriving there on 16th September. The cruiser *Flora* escorted them to Aden, Sailing on 14th October and arriving there on 26th. The cruiser *Marathon* escorted them on from Aden. In 1913, when these vessels were put in classes, she became a member of the 'A' Class, but shortly afterwards, in 1916, she was sold at Hong Kong. Note the officer wearing a pith helmet and being rowed by two sailors in a boat marked with an Admiral's flag and fitted with an awning.

HMS ST ENOCH

9 September 1897

ALTHOUGH not appearing in the Navy List for 1897, the *St Enoch* is seen wearing the White Ensign as she enters Grand Harbour with a large crew of naval ratings—at least 20 can be seen in the photograph. This steam dredger had been Yard Craft No 26 at Portsmouth from 1894. When she was transferred to Malta, to become Yard Craft No 90 there she was taken by the *Royal Sovereign*, who provided the crew and acted as escort. They sailed from Portsmouth on 23 August, and put in at Ferrol and Gibraltar before arriving at Malta on 9th September. In 1899 she went on to Wei Hai Wei, being sold there in April 1930.

HMS VULCAN

In Dry Dock

IN these two photographs taken on separate occasions (1897 and 1899), the Torpedo Boat Carrier *Vulcan* is seen in Dry Dock in Malta. From the bow angle her large cranes for lifting in and out her six torpedo boats are prominent. These large cranes were hydraulically operated and were 65 ft high. They were supported by feet on castings at the ship's double bottoms. They could lift her 60 ft torpedo boats clear of the torpedo nets, thus avoiding becoming vulnerable while hoisting or lowering her boats. A similar arrangement was built into the battleships *Swiftsure* and *Triumph* when they were being built for the Chilean Navy. From both views the fine lines of the hull, designed to give her a speed of 20 knots, can be seen. However, when built at Portsmouth, being launched in June 1889, weaknesses in her hull were noted, and also her engines were not stretched on trials, for she reached just 18½ knots. In 1895 she carried out a 48 hour trial from Volo to Malta and averaged 17.75 knots. She also carried eight 4.7 inch guns and twelve 3 pounders and was fitted with two submerged and four above water torpedo tubes, so that she could defend herself while launching her torpedo boats against an enemy. On each side of her stern are large 'panama' fairleads and beside them are a couple of the modern 'Kisby' lifebuoys for use if a man went overboard in daylight. Her 4.7 inch guns were removed when she became a submarine depot ship in 1915, and by 1922 all her guns had been removed. She became *Defiance III* at Devonport in 1931 and was broken up in Belgium in December, 1955 after the Torpedo School at Devonport was moved ashore to Wilcove.

VULCAN

HMS HOOD

10 November 1897

SEEN here leaving Grand Harbour for a short cruise to Palermo and Naples, the *Hood* was a half sister to the *Royal Sovereign* Class, having her 13.5 inch guns built in turrets instead of in barbettes. This gave them a rather stubby look. Because of the extra weight of the turrets, she had to be cut down fore and aft, and as a result proved to be a poor seaboat in comparison with her half sisters, so served most of her active life in the calmer waters of the Mediterranean. The scroll work on her bow includes a painting of the Admiral's head, although on her stern is just her name in plain letters. A pair of sailors can be seen on the starboard bridge wing heaving the lead, an exposed and chilly task in colder waters and now replaced by echo sounders. Her anti-torpedo nets can be seen furled between her upper and lower 6 inch gun batteries, while a Royal Marine guard is mounted on the quarterdeck. She was launched at Chatham in July 1891, refitted there in 1900 and again in 1902 and 1903, on return from the Mediterranean for the last time. After a period in reserve and as receiving ship at Queenstown, she was removed from the effective list in 1911, was used as a target for torpedo tests and in 1914 was sunk as a blockship at the South entrance to Portland Harbour.

HMS VICTORIOUS

February 1898

THE *Victorious* was a *Majestic* Class Battleship, launched at Chatham in October 1895 and completed in November 1896. On trials she achieved 18.7 knots, and shared the honour of being the fastest of the class with the *Caesar* (see page 94). Having spent a short period in reserve, and then with the Channel Fleet, she commissioned in January 1898, and is seen here on her way out to the Far East. She arrived at Malta on 10th February, at Suez on 22nd, and at Aden on 7th March. She grounded twice while passing through the Suez Canal. Sailors can be seen suspended under the upper top of the mainmast painting. It was said of these ships that the masting was superb, but that two tops with protection would have been better than four without, but that the twelve 3 pounders in the tops would make an awful sweep of the upper deck of an enemy so long as the gunners held out alive. This reflects the short range at which naval actions were envisaged to take place at that time. *Victorious* had the older pear shaped barbettes, which had hoods made of ten inches of armour. She served again in the Mediterranean from 1900 to 1903, and then in the Channel Fleet and Atlantic Fleet. During the war she was sent to defend the Humber. Her main armament was removed in 1915 for use of the *Lord Clive* Class of monitors, the guns being given extra elevation to increase their range to 21,000 yards. She became the Base Ship for Scapa Flow dockyard workers from 1916 to 1919 and was sold in 1922, and resold in April, 1923 for breaking up.

HMS HARRIER

March 1898

THE *Harrier* is seen leaving Sliema Creek to cruise off Malta, presenting a rather strange appearance with her large poop and widely set funnels. Her port twin torpedo carriage can be seen in her waist, while her 4.7 inch guns are high on the forecastle and poop. These *Halcyon* (or *Dryad*) Class Torpedo Gunboats were the last of their type, for the new style destroyers were then being built and were proving far more effective for torpedo work. The *Harrier* had been launched at Devonport in February 1894 and completed in 1895, when she carried out trials, reaching 19 knots, half a knot better than her designed speed. She then went into reserve and commissioned two years later for the Mediterranean, where she stayed with only a short break until 1905. After a short spell of Fishery Protection duties, she ran for the Navigation School at Portsmouth until 1913, when she was converted for minesweeping duties. She formed part of the Downs Boarding Flotilla at the outbreak of the Great War. She was placed on the disposal list in 1919, sold to Venezualan agents in February 1920, but was arrested and placed in the custody of the Admiralty Marshal. In September 1921, she was auctioned to a Welsh firm who proposed to convert her for commercial use.

HMS ILLUSTRIOUS

Summer 1898

A *MAJESTIC* Class battleship of 14,900 tons, the *Illustrious* had been floated out of dock at Chatham in September 1896 and completed in April 1898. She sailed for the Mediterranean from Portsmouth on 21 June, arriving at Malta in September. The class were armed with 12 inch guns, which were superior to the previous battleships' 13.5 inch guns in all ways except for the weight of shell. They also introduced cordite instead of gunpowder as the propellant for the 12 inch shells and the secondary armament of 6 inch guns. This propellant was more effective, and much lighter allowing the 6 inch guns to have 50 per cent more ammunition per gun. *Illustrious*, and her sister the *Caesar* (see page 94) were unusual in the class in having round barbettes instead of pear shaped ones. This allowed for all round loading and also created a break in the ammunition supply, a great safety measure. Rated at 17 knots, the *Illustrious* only achieved 16½ knots on trials. Although the trials were run in bad weather, the *Illustrious* did have a lot of trouble with her machinery. Fitted with 9 inch armour on her sides amidships, the ends of the ship were lightly plated with ½ inch steel, and on one occasion a torpedo boat rammed her and holed her. To avoid this happening again, the class were later sheathed in wood. Shortly after this photograph was taken she took part in the Crete insurrection operations. She returned home in 1904 and served in the Channel Fleet till 1908. During the war she was a guardship, her 12 inch guns being removed in 1915 for use in the *Lord Clive* Class of monitors. In 1916 she was used as an ammunition store ship on the Tyne, and later was a store ship at Portsmouth. She was sold in June 1920.

GRAND HAROUR

November 1898

THE *Royal Sovereign* Class battleship *Ramillies* (see page 97) lies in the foreground, whilst the *Majestic* Class battleship *Caesar* (see page 94) can be seen over her bow. Though the *Majestics* were only marginally larger, their hooded barbettes gave them a more solid appearance, and the provision of the secondary armament in casemates and the fitting of four full tops on the masts gave them a very different look. A single funnel picket boat lies alongside the *Ramillies*, whilst a twin funnel one lies alongside the *Caesar*. The *Majestics* were fitted with four submerged and one above water 18 inch torpedo tubes, and carried 17 torpedoes for them. They also carried five 14 inch torpedoes for launching from their picket boats. *Ramillies* has ports open in her hull just forward of the forebridge and aft of the after bridge, revealing the searchlights fitted to illuminate any torpedo craft approaching her at night so that her 6 inch guns could be brought to bear. It was also hoped that the beam would dazzle any assailants. Over the stern of the *Ramillies* is a cruiser, whilst ahead of the *Caesar* is the German Imperial Yacht *Hohenzollern*, launched at Stettin in 1893 by the Vulcan Shipping Company. Fitted with three masts and two bell-mouth funnels, she was decorated with gilding, had a ram bow and turtle back forecastle. Her triple expansion engines gave her a speed of 21½ knots.

Previous pages

SHIPS IN HARBOUR

circa May 1896

IN this magnificent panorama of Grand Harbour looking towards French Creek, the *Barfleur* is the Flagship on the left (see page 73) wearing the flag of Rear Admiral R A Harris, temporarily whilst the *Trafalgar* was under refit. *Trafalgar* (see page 69) can be seen without her funnels in the dock by the tall chimney above the *Barfleur*'s stern. Here she is in Mediterranean livery. When in the Far East, she, and her sister, *Centurion*, were painted white, but *Barfleur*'s had a blue tinge, so the sailors called her the '*Farbluer*'. Beyond her lies an *Astraea* Class cruiser. The battleship in the right foreground is the *Admiral* Class battleship *Camperdown* (see page 48), carrying her new large signal mast forward of her funnels. Over her bow can be seen the *Nile* (see page 76), both these vessels nearing the end of their Mediterranean days, whilst a *Royal Sovereign* Class battleship (probably *Ramillies*) can be seen over her stern wearing the Commander-in-Chief's flag. Refitting by the crane can be seen the *Collingwood* (see page 54) and the *Admiral* Class battleship *Rodney*.

HMS HUSSAR

August 1899

THE Torpedo Gunboat *Hussar* was built at Devonport, being completed in January 1895. She was the fastest of her class (the *Halcyon* Class) on trials at 19.7 knots with the least horsepower (3553). Her starboard twin revolving torpedo tubes can be seen here as she leaves Sliema Creek, with sailors heaving the lead on both bridge wings, but with most of the ship's company mustered in the starboard waist. She had returned to Malta from Tunis on 7 August and was sailing again with her sisters *Harrier* and *Hazard* to join the cruising Squadron at Alcudia Bay (Majorca). She served in the Mediterranean from 1896 to 1905 with a couple of short breaks, then was used as an RNR Drill Ship and later was employed on Fishery Protection duties. Her guns were removed in 1907, and she returned to the Mediterranean to become the Admiral's yacht, but her forward gun was replaced in 1909 for patrol work. Converted to a minesweeper in 1914, she served in the Dardanelles, where her Commanding Officer, Commander E Unwin, a midshipman and two seamen won Victoria Crosses. After the war, *Hussar* returned to her duties as yacht and despatch vessel for the Commander-in-Chief, Mediterranean. She was sold in December 1920, and resold in July 1921 for breaking up in Malta.

HMS CAESAR

23 May 1900

IN this view the close, side by side arrangements of this class's funnels can be seen clearly. The next class, the *Canopus* Class, had twin funnels fitted fore and aft. Her sailors are in half blues as she leaves harbour making a fair amount of smoke as she heads for Alcudia Bay to join the Fleet for a cruise. She, with her sister, the *Illustrious* (see page 89), was fitted with round barbettes. She also had her bridge fitted forward over the conning tower, and not under the mast. *Illustrious* was similar, whilst in *Victorious* (see page 87) the bridge was under the mast. Built at Portsmouth and completed in January 1898, she commissioned for the Mediterranean, but served for a while in the Channel Squadron before arriving at Malta in May 1898. She was refitted at Malta later in 1900, and returned to Portsmouth in 1904. On joining the Channel Fleet she became a Flagship. She was refitted at Devonport in 1907 when a fire control system was fitted. She served with the Third and Fourth Divisions of the Home Fleet up to the outbreak of the war, when she commissioned for Channel operations and she operated in the Atlantic and on the North American and West Indies Station. In 1918 and 1919 she served in the Mediterranean and Black Sea. She was sold in November 1921 and was taken to Germany in July 1922 for breaking up.

SAILORS LANDING AT MALTA

THIS undated photograph shows the sailors' uniform at the turn of the century. White tunics for officers were not introduced until 1885 and the shoulder boards for them not until 1891. The Petty Officers' badge can be seen, introduced for First Class Petty Officers in 1857, and there are sennet hats, which lingered until 1922, when they were finally abolished because of the difficulty in stowing them in modern vessels. Until then, ships of any size were fitted with hatracks overhead on the messdecks and were supplied with a hat block and gelatine resin for reshaping and stiffening the straw when the hats lost their shape. This photograph appears to be taken in the Marsa Creek area, with flat land visible on the left and the buoys for ships on the right. Note the upended 32 pounder cannon used as a bollard. The boats in the foreground and one in the background are of the collapsible type, made of canvas and wood, and used in vessels such as the early torpedo boat destroyers where deck space was at a premium.

DOCKYARD CREEK

1900

AT the head of the left hand trot in this picture is the Third Class cruiser *Scout* (see page 100) resplendent in white livery, whilst astern of her lies an *Astraea* Class cruiser (see page 80). Next is the only three funnel vessel of the *Alarm* Class of torpedo gunboats, the *Speedy*. She was fitted with Thornycroft water tube boilers and was the only one of the class to achieve 20 knots. Completed by Thornycroft in February 1894, she was lost when mined off the Humber in September 1914. Astern of her lies either the *Bellona* or *Barham*, three masted, two funnel cruisers of 1890-91 (see page 60) and next is another early third class cruiser, possibly the *Surprise* (1885) which was fitted as a despatch vessel with accommodation for senior officers. Last in the line is the Torpedo Gunboat *Hebe* of the *Alarm* Class, launched at Sheerness in June 1892. She became a submarine depot ship in 1910 and was broken up in 1920. It is interesting to compare her with her three funnelled sister further up the trot. On the other side of the creek lies the *Hibernia* (see page 28), and lying astern of her is a four funnel destroyer of the type built by Cammell Lairds from 1896-1901, capable of 30 knots. This view provides a real mixture of the ancient and the modern.

HMS RAMILLIES

14 August 1901

THIS *Royal Sovereign* Class battleship was launched on the Clyde in 1892 and had commissioned at Portsmouth in 1893 for service in the Mediterranean as Flagship of the Rear Admiral Mediterranean. She is seen here entering harbour wearing the flag of The Lord Charles W D Beresford CB returning from a cruise in the Levant. A year after she joined the Fleet she had bilge keels fitted, at Malta, which improved her ability to maintain speed in a seaway. This class was the prototype for all the pre-Dreadnought battleships. Her 13.5 inch guns can be seen here in an elevated position, not having the protection for the crew that was introduced with the *Renown* (see page 98) . They could be loaded only in the centreline position, a limitation that was corrected in later classes. The ship's company can be seen on the upper deck wearing their sennet hats, and the cutter is manned for the buoy jumpers. She has large windscoops rigged between the bridge and funnels, and the ports for two of her seven torpedo tubes can be seen just above the waterline below the forward and after bridges. She has yet to hoist her harbour ensign. She was sold in 1913.

DESTROYERS

April 1901

A GROUP of destroyers in their Mediterranean colours are seen escorting the *Renown* (see this page and opposite) off Malta on their return from exercises off Corfu. The early destroyers were not listed by classes until 1913, when a more standard system of building was brought into force. Admiral Fisher, who was flying his flag in the *Renown*, commented the year before that the destroyers 'were dashing about like mad and in the middle of it all and torpedoing everyone'. He also, however, criticised the extra spares and stores that the depot ship *Tyne* had to carry because of the lack of standardisation in design. The differing turning circles and speeds also made fleetwork difficult. The nearer vessels are Thornycroft 30 knotters of 1896-8. They became the D Class, and were built of a new high tensile steel to save weight. Included here are the *Coquette*, *Cygnet* and *Foam*. Beyond are Laird built 30 knotters with four funnels (including *Orwell*, *Earnest* and *Griffon*). The four funnel destroyers were placed in the B Class in 1913.

HMS RENOWN

A BATTLESHIP of an enlarged *Centurion* Class design, she was built with 10 inch guns as her designed 12 inch guns were not ready in time. Her guns could elevate to 35 degrees, and they were mounted in protective shields—a novel feature at the time. The shields can be seen clearly in this picture. Launched at Pembroke Dock in 1895, she was fitted out at Devonport, completing in 1897. She then went to the North America and West Indies Station where she became the Flagship of Admiral Fisher. When Admiral Fisher became the Commander-in-Chief of the Mediterranean Station in 1899, he had the *Renown* transferred to be his Flagship again. Although not wearing Fisher's flag in the picture opposite (top) (March 1901) the boats are marked with it. Her captain was The Hon Hugh Tyrwhitt. Her cutter is manned ready for dropping the buoy jumpers on entering harbour. In the photograph above she is seen flying Admiral Fisher's flag as she leaves Grand Harbour (June 1902), with his second in command in the *Ramillies* astern, and the *Implacable* (see page 124) to starboard. Later she was fitted out for Royal Tours of India. She had an excellent reputation as a seaboat and steamer. She became a Stokers' Training ship at Portsmouth in 1909 and was sold for breaking up at Blyth in 1914.

HMS SCOUT

August 1901

A TORPEDO cruiser (sister to the *Fearless* (see page 56)) built on the Clyde by Thomsons in 1885, her speed of 17 knots was too slow for her intended role of defending the Fleet against attack by torpedo boats, and she also proved to be a poor seaboat, so she was employed as a slow cruiser for commerce protection. Originally she had been fitted with sails, but these were removed at the same time as her 5 inch guns were replaced by 4.7 inch quick firing guns. During her second commission, which started on 3 October 1889, she was commanded by HRH Prince Louis of Battenburg GCB. She served in the Mediterranean from 1887 to 1903, and was awarded the Battle Honour 'Gambia' in the year of this photograph. Her crew were in the naval brigades in South Africa (1900-1901). She is seen here in her white colouring, having arrived at Malta from Mombasa for docking earlier in the month. Her Commanding Officer had changed two days prior to this photograph. She left Malta on 5 October for the Red Sea via Suda Bay and Alexandria. She was sold in 1904.

HMS ANDROMEDA
September 1901

A FIRST class cruiser, built at Pembroke Dock, being completed two years before this photograph was taken, she was one of the *Diadem* Class, of sound design but slow (20 knots) and lacking any heavy armament. This class carried sixteen 6 inch guns with five guns on each side and paired mountings on the forecastle and quarterdeck. They were made obsolete when armoured cruisers were re-introduced, and they spent much of their life in reserve. Seen here approaching Grand Harbour, she has her torpedo nets stowed and the supporting booms can be seen lying against her side. Her cutter is manned and the hands are preparing to fall in for entering harbour. In 1902 she was the Flagship of the Rear Admiral of the Mediterranean Cruiser Squadron. She became the *Powerful II* (1913), the *Impregnable II* (1919) and the *Defiance* (1931) and was finally broken up in Belgium in 1956.

Overleaf

HMS ROYAL SOVEREIGN
June 1902

NAMESHIP of her class, the *Royal Sovereign* was floated out of dock at Portsmouth on 26 February 1891, being christened by Queen Victoria. She was completed in May 1892, and was the first armoured ship to achieve 18 knots when run deep on trials. Her engines developed 13,360 Horse Power, over 2000 more than her sister ships. Her sister *Ramillies* did run at 18.27 knots on trials, but at light load. These ships were built with a high freeboard, and the sides were given a tumble home, reminiscent of wooden warships, to save weight. Their high bow and tall sides enabled them to maintain speed in heavy seas. The *Royal Sovereign* joined the Mediterranean Fleet in June 1897. In May 1899 she recommissioned for another spell in the Mediterranean with the officers and crew from the *Trafalgar* (see page 69). In November 1901, a 6 inch gun exploded killing six men and injuring another 19. On being relieved in the Mediterranean by the *London* in August 1902, she undertook Port Guard duties at Portsmouth. During a refit in 1903 casemates were fitted to her upper 6 inch guns in lieu of shields. After a period in reserve she commissioned with a skeleton crew from 1909 to 1913. She was then replaced in reserve and was sold in October 1913 for breaking up at Genoa. The larger distance apart of her funnels than *Renown*'s can be seen in this photograph (cf page 99).

CHANNEL FLEET

1902

THE Channel Fleet is seen in Grand Harbour in September 1902, having also visited the Mediterranean earlier in the year (March-April). They returned in the August, leaving Gibraltar on the 28th for Malta. They fill the harbour, there being few Mediterranean Fleet representatives. The six *Majestic* Class battleships (*Majestic* - Flagship of Admiral Sir Arthur Wilson KCB VC; *Magnificent* - second flagship; *Mars*; *Hannibal*; *Jupiter* and *Prince George*) are split between Bighi Bay and the area off Corradino. The *St George*, *Furious* and a *Pelorus* Class cruiser share Dockyard Creek with the *Hibernia* and destroyers, whilst there is a line of cruisers in the trot towards Marsa Creek, led by the *Sutlej* (1899), followed by an *Eclipse* Class (*Doris*, *Juno* or *Minerva*), then the *Diadem* Class *Niobe* (sold to Canada in 1910) with another *Eclipse* Class astern of her.

The FLEET OUTSIDE HARBOUR

The Mediterranean Fleet is seen here off Grand Harbour, arriving from Sardinia. The *Renown* leads the line as almost her last act as Flagship of Admiral Fisher (see page 98) whilst the *Ramillies* is astern of her. The newly completed *Aboukir* (see page 128) with her four funnels seems out of place amongst the heavy twin funnel battleships—but at 12,000 tons she was only 350 tons lighter than *Renown* and she was 50 feet longer than even the *Majestic* Class battleships.

HMS BRUIZER

12 May 1902

BRUIZER was one of the 1893/4 initial order of torpedo boat destroyers from Thornycroft at Chiswick. Of 265 tons, she was launched on 27 February 1895 and completed in June of that year. Designed as a 27 knotter, she achieved 27.97 knots on trials. Her two propellers were driven by Thornycroft water tube boilers. Note the typical Thornycroft stern. Originally these boats were to have a single bow torpedo tube, and none on deck, but the early boats found that the bow tube made the bridge extremely wet and the risk existed of the boat overrunning its own torpedo when fired, so two tubes were mounted on the deck instead. Here she is just arriving from Sardinia with the Fleet. She was under the command of Lt and Cdr R C Kemble Lambert and was a tender to the *Canopus* (see page 113). She was sold in May 1914, to Cashmore of Newport, although many of her sisters were retained a little longer, and so served through the war, one being lost in a collision and another by mining.

HMS LOCUST

22 August 1902

A LAIRD 30 knotter, the *Locust* was launched in December 1896, and completed in July 1898. Her engines developed 6000 horse power, and her twin screws gave her 30.16 knots on trials. Sturdily built at 355 tons, she was not a very manoeuvrable vessel. Her armament was a single 12 pounder gun and five 6 pounder guns. She was fitted with two torpedo tubes. A collapsible boat is covered in black canvas abreast the forefunnel. This type of boat can be seen in use on page 95. Here the *Locust* is seen when a tender to the *Orion* (see page 36) under the command of Lieutenant and Commander A B S Dutton. She is preparing to work with Mediterranean and Channel Fleets in the Western Mediterranean. She survived the First World War and was sold in 1919.

HMS BULWARK

21 May 1902

THE *Bulwark* was one of the *London* Class battleships, a virtual repeat of the *Formidable* Class in order to allow more time for work on the next battleship design (the *Duncan* Class). *Bulwark* was launched at Devonport on 18 October 1899 and was completed in March 1902. She then proceeded to the Mediterranean to relieve the *Royal Oak* and is seen here sailing to Terranova and Genoa to bring Admiral Sir Compton E Domvile KCB to Malta. He was the Commander-in-Chief and he hoisted his flag in her in June. She was later the Flagship of Admiral Lord Charles Beresford who was the next Commander-in-Chief. In 1907 she returned home, and in 1908 Captain R F Scott took command, the most junior Captain in command of a battleship at the time. At the outbreak of war she was in the Fifth Battle Squadron in the Channel. On 26th November 1914, while embarking ammunition in the Medway, she blew up. Only 12 out of a ship's company of over 700 were saved. A Court of Enquiry estimated that the explosion had been caused by the accidental ignition of her ammunition.

HMS VENGEANCE

May 1902

KNOWN to her sailors as 'The Lord's Own', the *Vengeance* was the first battleship to be built, armoured, gunned and engined all by the same firm—Vickers at Barrow. A *Canopus* Class battleship of 12,950 tons, she was launched on 23 August 1898, and was completed in April 1902. The class were armed with two twin 12 inch mountings. *Vengeance* was the only one of the class whose gun houses were of Krupp steel, and they were flat sided instead of curved. She was also fitted with Vickers chain rammers for loading the 12 inch at any elevation, thus reducing the firing interval from 45 to 32 seconds. This class was built without forced draught, as forcing had proved hard on boilers, but could achieve 18 knots at natural draught (*Ocean* achieving 18.5 knots). *Vengeance* was considered to be the best steamer of all the class. In April 1902, on completion, the *Vengeance* joined the Mediterranean Fleet arriving at Malta on 13 May and from appearances (anchor cable broken, boat lowered at an angle and 'maypole' on her quarterdeck) she seems to be conducting evolutions. In June 1903 she sailed for China, where she remained for two years. She then had a period in reserve and refit and was gunnery training ship at Chatham in 1909 and the Nore 1913-14. At the outbreak of war she was in the Channel, then served in the Atlantic, Cameroons, at Egypt, then the Cape Verde Station prior to being sent to the Dardanelles as Admiral de Robeck's flagship. Later she was guardship in Egypt, and served in the East Indies and the Cape Station before becoming an ordnance depot at Devonport in 1918. She was sold in 1921 and was broken up at Dover in 1923.

HMS IRRESISTIBLE

December 1902

SEEN here in the new grey paint scheme, the *Irresistible* was a *Formidable* Class battleship, completed by Chatham in February 1902. The class had inward turning propellers which gave them added speed, but made them very unhandy when manoeuvring at low speeds. Her heavier 12 inch guns could pierce armour at 50% greater range than the previous 12 inch gun. On completion she commissioned for the Mediterranean. She is seen here in Marsamxett Creek prior to sailing for Astacos with the *Bulwark* (see page 109) and *Pioneer*. She remained in the Mediterranean till 1908, when she joined the Channel Squadron. At the outbreak of the war she was engaged in the landing of marines at Ostend, and in February 1915 she joined the Dardanelles force. On 18th March, at 3.14 pm, she fell out of line whilst heading for the narrows when a shell exploded alongside her. She had just regained her position when she set off a mine under her starboard engine room, which flooded rapidly, the midship bulkhead crumpled and the port engine room flooded too. The destroyer *Wear* closed and took off the crew whilst under fire from ashore. Ten men were left onboard to take a towing wire from the *Ocean*, but at 5.50 they abandoned her and the *Ocean* was mined on her way back. She sank at about 7.30, and the *Ocean* sank 3 hours later.

HMS FORMIDABLE

February 1903

NAMESHIP of a new class of battleships, the *Formidable* was launched at Portsmouth in November 1898. The ships were 1500 tons heavier than the preceeding *Canopus* Class, mainly because they carried longer and heavier 12 inch guns. They had the same fore and aft funnel arrangement, but with the funnels set wider apart. Completed in September 1901, after delays caused by the liquidation of the firm building her engines, she commissioned for the Mediterranean in the October with officers and men from the *Resolution*. Before leaving for the Mediterranean she carried out trials of the new stockless anchor, and rode out a storm at Spithead on one anchor only. Despite this, the new anchor was not introduced until after the *Duncan* Class had been designed, and she is seen here with the old pattern anchor, and in grey paintwork. She remained in the Mediterranean until 1908, and then was refitted and served with the Home and Atlantic Fleets. At the outbreak of the war she took the Portsmouth battalion to Ostend (August) and then was with the 5th Battle Squadron at Sheerness. The 5th Squadron was relieved by the 6th Battle Squadron in December so that they could go to Portland to exercise their gunnery. The Squadron arrived off Portland on 31st and it was decided that they would stay at sea overnight. At 2.30 am on 1st, half an hour before the Squadron were to turn East off Start Point, the *Formidable*, the last ship in the line, was torpedoed abreast the forefunnel by *U24* (Lt Cdr Schneider). Half an hour later she was hit again, on the opposite side abreast the after funnel. At 4.45 am she lurched, heeled over rapidly and sank. 547 were lost out of her crew of 780.

HMS CANOPUS

29 December 1902

THE *Canopus* is seen here sailing for Gibraltar with *Victorious, Diana* and *Bacchante*. The nameship of a class of battleships which re-introduced the fore and aft funnel arrangement, also being the first British battleships with water tube boilers. The after funnel was as long fore and aft as the forefunnel, but was wider. This class also introduced Krupp Cemented armour, allowing a considerable saving in weight (1 inch of Krupp armour was as effective as 2.6 inches of iron). They had very strong, prominent rams with a cast steel stempiece, the framework being filled with cork to make it resilient. The *Ocean* hit the breakwater at Portland with no damage to herself in 1902. Another innovation was the large sponsons for the forward and after 6 inch guns to give them end on arcs of fire. Note here the unusual colour scheme of her hull. The *Canopus* completed at Portsmouth in December 1899, and served in the Mediterranean from then until April 1903, though most of her sisters were on the China Station. She commissioned for China in 1905, but joined the Atlantic Fleet instead. At the outbreak of war she was guardship at St Vincent, and was sent to join Admiral Cradock's Squadron, but was 250 miles away when that Squadron was engaged at Coronel. Ordered to be guardship at Port Stanley, her guns were the first to engage Von Spee's Squadron at the Battle of the Falkland Islands. In early 1915 she was in action again at the Dardanelles. She returned home in 1916 and became an accommodation ship at Chatham, being sold in 1920.

HMS MONTAGU

7 September 1903

THE *Montagu* was one of the six *Duncan* Class battleships built to provide faster battleships than the Russians were building. They could steam at 19 knots, and carried the same armament as the *Formidable* Classes, but were slightly longer (by 5ft) and 6 inches more in the beam. They were, however, lighter by 1000 tons, mainly achieved through savings in armour and protection. None of class had wind cowls, and *Montagu* and the *Albemarle* were fitted with stockless anchors, whilst the rest retained the older pattern anchors. Overall, with their even funnels and open tertiary batteries amidships, they were better looking than the *Formidables* though less well protected. She was completed at Devonport in July 1903, and commissioned for the Mediterranean, where the whole class served at the start of their careers. She is seen here on the day she first arrived at Malta. In 1905 she joined the Channel Fleet. On 30 May 1906, she struck the Shutters on the South Western end of Lundy Island in a fog, whilst carrying out experiments on devices designed to prevent wireless being tapped. Her sister ship, the *Duncan* (see page 117) grounded whilst trying to tow her off, and had be repaired herself. In August the salvage operation was abandoned, her 12 inch guns were removed and taken to Pembroke Dock and then shipped to Woolwich, whilst her torpedo tubes were destroyed with explosives. The wreck was sold in February 1907. They were last British battleships to wear the black, white and buff livery before the Navy changed to uniform grey.

HMS DUNCAN

November 1903

THE *Duncan* was nameship of her class, and was built at the Thames Iron Works, being launched on 21 March 1901. This was the first class of battleships for the Navy to cost over one million pounds each, the average cost was £1,093,000. She commissioned for the Mediterranean on 8 October 1903, at Chatham under Captain Henry B Jackson—already a Fellow of the Royal Society and later an Admiral of the Fleet, he had been at the forefront of wireless invention. The *Duncan* arrived at Malta for the first time on 8 November and sailed for Platea on 27th. She stayed in the Mediterranean until 1905. In July 1906 she grounded while trying to assist her sister the *Montagu* aground off Lundy (see page 114) and repairs took 72 days. She returned to the Mediterranean in 1908 after a refit in Gibraltar and remained there till 1912. At the outbreak of war she was employed on the Northern Patrol and in late 1914 she joined the 3rd Battle Squadron at Portland and Dover. She went back to the Mediterranean in 1915, returning home in 1917 and was placed in reserve, being sold in 1920.

Opposite

HMS DIANA

March 1903

A SECOND class cruiser of the *Eclipse* Class, *Diana* had commissioned for the Mediterranean at Chatham on 15 January 1901 after a spell of trooping duties to Australia. Although completed by Fairfields in June 1897, her only other service prior to this had been when she commissioned for the Diamond Jubilee Review. This class were improved *Astraeas*, with an extended turtle back forecastle for sea-worthiness. They were the first British cruisers to be fitted with tops. In this month she temporarily carried the flag of the Rear Admiral commanding the Cruiser Division, it having transferred from the *Bacchante*. Soon afterwards the flag transferred again, this time to the *Aboukir* (see page 122). In 1904 she paid off and underwent a refit in which her mixed armament of five 6 inch guns and six 4.7 inch guns was replaced by a homogenous arrangement of eleven 6 inch guns. She then returned to the Mediterranean station until 1913, with a short period in 1909 when she was detached to the East Indies. In August 1908 she relieved the *Argonaut* as the parent ship for TBDs in the Mediterranean. During the war she served with the 12th Cruiser Squadron in the Western Channel area, then, after a refit, went out to the China Station and then the East Indies. She paid off in June 1919, and was sold in 1920. Her signal department can be seen here, a cheerful mixture of 'sparkers' and 'bunting tossers', with the Petty Officer Telegraphist holding their mascot.

HMS QUEEN

April 1904

THE *Queen* and the *Prince of Wales* were second repeats of the *Formidable* Class whilst designs for the *Duncan* Class were being prepared. She was launched by Queen Alexandra at Devonport on 8 March 1902—King Edward VII laying down the battleship of his name on the same occasion. The repeat *Formidables* had no lower deck scuttles forward, and these last pair again differed in having open 12 pounder batteries amidships, a reflection of the argument as to whether the protective screens provided cover or served to explode passing shells. She has windsails instead of cowls. Stockless anchors were first fitted in these ships and they were the last of the series of battleships with the wider after funnel. The *Queen* could be identified by the deckhouse mounted on her after bridge. She first commissioned on 7th April 1904 for the Mediterranean and served there until 1908, being Fleet Flagship in 1907. After a refit she served in the Atlantic Fleet. At the opening of the war she operated off the Belgium coast and the Channel. In March 1915 she and the *Implacable* were sent to the Dardanelles, and covered the Anzac landings in April. She then went with *Venerable* (see page 125) to the Adriatic, in support of the Italian forces. She was Base Ship at Taranto supporting the Anti-submarine Net Barrage Flotilla from 1916 to 1918. In 1917 it was agreed that her large guns should be dismounted and used as heavy artillery on the land front to assist the Italians. She returned to England to pay off in 1919 and was sold in 1920, her hull being broken up at Preston in 1921.

TORPEDO BOAT No 22

1904

A SISTER to Torpedo Boat No 21 (see page 41) she was built by Thornycroft and launched in 1895. Their prototype, the *Childers* travelled out to Australia under her own steam. With two torpedo tubes and two guns, and a crew of 14, she displaced just 64 tons and when built could reach 20 knots, but in her later days could barely manage 15 knots. She is seen here entering Grand Harbour, Ricasoli Fort and Rinella Creek can be seen beyond her. She was sold at Malta in 1907.

HMS DRAKE

8 June 1904

THE nameship of a class of Armoured cruisers, the *Drake* is seen here arriving at Malta from Portsmouth, from where she had sailed on 2nd May. Two days later she left Malta to join the cruiser squadron at Ajaccio. Her Commanding Officer was Captain J R Jellicoe CB, who had taken command in the August of the year before, and who was relieved by F C D Sturdee in November 1904. She became Flagship of Rear Admiral HSH Prince Louis of Battenburg in 1905. Enlarged *Cressys*, with more powerful machinery, they were among the fastest ships in the world when they were completed and often exceeded their designed speed (23 kts) during service. *Drake* was the fastest on trials at 24.11 knots. The class had no ventilation cowls, and deck fittings had been reduced to the minimum to reduce the risk of fire, giving them a very clean look. During the war she served with the Grand Fleet until 1915. After a refit she joined the North America and West Indies Station and was employed escorting Atlantic Convoys when they were instituted. On 2 October 1917 just after she had dispersed convoy HH24 from Hampton Roads off the North Coast of Ireland, she was torpedoed by *U79*. Her Captain found she could still steam, so headed for Rathlin Island and called for assistance, 8 destroyers and 4 sloops escorted her into Church Bay where she anchored. She then started to heel over, and had to be abandoned. She capsized and sank whilst still at anchor. 19 men were lost. One of her anchors was recovered by divers and is now mounted in the present *Drake*, the barracks at Devonport.

HMS ESSEX

March 1905

PART of the Second Cruiser Squadron of the Atlantic Fleet, the *Essex* is seen here at Malta during a spell in the Mediterranean, which included visits to Malaga, Naples, Marmarice and Gibraltar. She had commissioned for the Second Cruiser Squadron in March 1904, at Devonport, having been completed that month at Pembroke Dock. A *Monmouth* Class Armoured Cruiser, the class were distinguished by their three (instead of the usual four) funnels, and individual identification was helped by the differing funnel steam pipe arrangements, *Essex* having pipes aft on each of her funnels. This class were really second class *Drake*s, with an all 6 inch gun armament, having a novel electrically operated twin 6 inch turret forward. This turret, however, was heavy and caused them to pitch in a seaway. *Essex* had been laid down on 1st January 1900. She was with the 4th Cruiser Squadron on the North America and West Indies Station when the war started, and captured a German merchant ship on 10th August 1914. She captured another a month later and a third in May 1916. She became a destroyer depot ship in 1916 and finally paid off at Devonport in 1919, being sold in 1920.

HMS ABOUKIR

20 March 1905

ABOUKIR was a *Cressy* Class Armoured Cruiser of 12,000 tons, armed with two 9.2 inch guns and twelve 6 inch guns. They were, in fact, 'belted *Diadems*' and were the first cruisers fitted with side armour since 1886. They were also the first cruisers designed for foreign service that did not have their hulls sheathed and coppered, a reflection of the improved docking facilities that were being provided abroad at the time. Seen here at Malta, with the cruiser *Essex* (see page 121) she sailed on 22nd for Suda Bay to join the Fleet. She had completed at Fairfields in April 1902, and immediately commissioned for the Mediterranean, and was about to return home. After a spell with a nucleus crew, she returned to the Mediterranean in 1907 and remained there till 1912. At the outbreak of war she was part of the 7th Cruiser Squadron operating in the North Sea. She was at the Battle of Heligoland Bight on 28 August 1914. In September 1914 she was patrolling an area (the 'Broad Fourteens') off the Dutch Coast. She was operating with her sister ships the *Hogue* and *Cressy* steaming in line abreast at barely 10 knots and not zig-zagging. At 6.30 am on 22nd she was torpedoed by *U9* (Weddigan) on her starboard side and after 25 minutes she rolled over and floated bottom up. Believing she had been mined, *Hogue* closed to rescue survivors and was hit by two torpedoes and sank in 10 minutes. The *Cressy* was then hit at 7.17 and again 15 minutes later. 60 officers and 1400 men were lost from all three ships, whilst 60 officers and 777 men were saved mainly by the Dutch steamships *Flora* and *Tritan* and two Lowestoft sailing trawlers.

HMS LEVIATHAN

30 May 1905

A SISTER ship to the *Drake* (see page 120) she is seen here wearing the flag of Rear Admiral The Hon Hedworth Lambton CVO, CB, after arriving from Suda Bay with the *Lancaster*. She had been completed by John Brown in June 1903. The class generally were good seaboats and carried an improved 9.2 inch gun on the forecastle and quarterdeck, with sixteen 6 inch guns along the sides. After returning home from the Mediterranean in 1907 she was refitted, then spent the rest of her service between the Home Fleet and the North America and West Indies Station. Most of her war was spent on the latter station and she was employed escorting Atlantic convoys in 1918. Shortly after the war she paid off and was sold to Hughes Bolckow, Blyth, for breaking up in March 1920. An *Eclipse* Class cruiser can be seen manoeuvring astern of her in this view.

HM Ships IMPLACABLE and MONMOUTH

April 1906

THE *Implacable* was a *Formidable* Class battleship, similar to her sisters *Formidable* and *Irresistible* (see pages 112 and 111). Launched at Devonport on 11th March 1899, she first commissioned on 10 September 1901 under Captain Prince Louis of Battenburg GCB, GCVO and served in the Mediterranean. She paid off at Devonport in 1904, but recommissioned for the Mediterranean again. In July 1905 she suffered a boiler room explosion in which 7 men died, and four months after this picture was taken she had another boiler explosion. In 1908 she paid off at Devonport and then refitted at Chatham. She returned to the Mediterranean in 1915 for the Dardanelles campaign and remained there until 1917, being the last British battleship in full commission in the Mediterranean during the war. She then served on Northern patrol and paid off in 1919, becoming a tender at Portland, and was sold in 1921 and was broken up in Germany, the only one of her class to survive the war.

The *Monmouth* as a sister to the *Essex* (see page 121) being completed at Glasgow in December 1903. She had completed her first commission with the Channel Fleet in April 1906 and recommissioned at Devonport a day later for the China Station. She is seen here on passage, arriving at Malta on 19th April and sailing on 21st. She reached Suez on 25th and then headed for Singapore. Her steam pipes were prominent, and were fitted forward of each funnel, with a fourth pipe on the after side of the after funnel. Sadly she was lost with all hands during the action off Coronel on 1st November 1914, being sunk by the *Nurnberg* at 9.18 pm with her flags flying.

TORPEDO BOATS

THIS group of torpedo boats, dressed overall, are of two distinct types. The nearer group of three are Yarrow built vessels, the nearest on being Number *063*, which would indicate the photograph was taken after 1906. In that year the early torpedo boats had a zero placed before their numbers, and the new coastal destroyers (oily wads), which were to have been named, were given the old numbers. TB *63* was launched in 1886 and was sold at Malta on 18th December 1919. She was of 75 tons, carried two 3 pounder guns and 5 torpedo tubes and could steam at 20 knots. Yarrow built boats had a curve in the decking to give more headroom and stowage space below. The two slightly larger boats beyond are Thornycroft vessels, with their distinctive sterns designed to keep the propellors in the water even when the boats were in rough seas. A contemporary poet wrote of these small vessels:

Only a number, not even a name
'How shall posterity hear of my fame
Perchance it may still live, after the grave
In the name of an ironclad under the wave'

To the right can be seen a Thornycroft two funnel 30 knot destroyer.

Overleaf

HMS VENERABLE

22 September 1903

THE *Venerable* was a sister ship of the *Bulwark* (see page 109) and was launched at Chatham in August 1899. Both these ships had ventilation cowls, whilst the third member of the class, the *London*, had windsails instead. She commissioned for the Mediterranean on completion, and became the Rear Admiral Custance's flagship on arrival there in November 1902. She is seen here returning from a visit to Barcelona. In June 1905 she touched the bottom in Algiers harbour and was slightly damaged. In 1906 she was fitted with a fire control system and in 1909 returned to Chatham for a refit. At the outbreak of the war she was employed in operations off the Belgian coast and in March 1915 she bombarded the battery at Westende. In May 1915, she was sent to the Dardanelles to replace the more modern *Queen Elizabeth* and operated off Suvla in the August. She then went to the Adriatic with three sister ships to contain the Austrian Fleet. She was sold in 1920 and was broken up in Germany. Her nickname was 'The Archdeacon'.

KING'S BIRTHDAY

November 1906

THE fleet lies in Grand Harbour firing a salute on the King's birthday, with a *Formidable* Class battleship (*Formidable* or *Implacable*) on the left. She has been fitted with the extra top on her foremast for fire control (compare with page 118). Royal Marines line both her fore and after bridges and sailors man her quarterdeck. On the right lies the *Carnarvon* (see page 129).

HMS CARNARVON

4 June 1906

THE *Carnarvon* had arrived from Piraeus on 25th May and sailed shortly afterwards to arrive at Gibraltar on 21st June, returning to Malta in July. At this time she was commanded by Captain Sir George J S Warrender Bt CB MVO, having commissioned in May 1905, at Devonport for the Mediterranean after completion by Beardmore. A *Devonshire* Class armoured cruiser, she had four 7.5 inch guns in single turrets, one forward, one aft and one each side of the bridge. The class were fitted with mixed boilers to try out various designs. She had 6 cylindrical boilers and 17 Niclause water tube boilers. Although all the class exceeded their designed 23 knots on trials the boilers proved unreliable. She returned home in 1907. She was Rear Admiral Stoddart's flagship (5th Cruiser Squadron) at the outbreak of the war, captured a German merchant ship on 24 August 1914 and was present at the Falklands action, when she helped sink the *Gneisnau*. She was undamaged in that action, and later helped in the search for the *Dresden*. On 22nd February 1915 she left Abrolhos Rocks heading north, but hit an uncharted rock an had to be beached with a 95 ft rip in her bottom. She was repaired at Rio de Janeiro. She spent the rest of the war on the North America and West Indies Station, covering Atlantic Convoys in 1917. She was a Cadets' Training Ship from 1919-21 and was sold in November 1921, being broken up in Germany.

INDEX

Named ships, with launch date in brackets, showing page number